Solid Mechanics and Its Applications

Founding Editor

G. M. L. Gladwell

Volume 269

Series Editors

J. R. Barber, Department of Mechanical Engineering, University of Michigan, Ann Arbor, MI, USA

Anders Klarbring, Mechanical Engineering, Linköping University, Linköping, Sweden

The fundamental questions arising in mechanics are: Why?, How?, and How much? The aim of this series is to provide lucid accounts written by authoritative researchers giving vision and insight in answering these questions on the subject of mechanics as it relates to solids. The scope of the series covers the entire spectrum of solid mechanics. Thus it includes the foundation of mechanics; variational formulations; computational mechanics; statics, kinematics and dynamics of rigid and elastic bodies; vibrations of solids and structures; dynamical systems and chaos; the theories of elasticity, plasticity and viscoelasticity; composite materials; rods, beams, shells and membranes; structural control and stability; soils, rocks and geomechanics; fracture; tribology; experimental mechanics; biomechanics and machine design. The median level of presentation is the first year graduate student. Some texts are monographs defining the current state of the field; others are accessible to final year undergraduates; but essentially the emphasis is on readability and clarity.

Springer and Professors Barber and Klarbring welcome book ideas from authors. Potential authors who wish to submit a book proposal should contact Dr. Mayra Castro, Senior Editor, Springer Heidelberg, Germany, email: mayra.castro@springer.com

Indexed by SCOPUS, Ei Compendex, EBSCO Discovery Service, OCLC, ProQuest Summon, Google Scholar and SpringerLink.

More information about this series at https://link.springer.com/bookseries/6557

Emmanuel E. Gdoutos

Experimental Mechanics

An Introduction

 Springer

Emmanuel E. Gdoutos ⓘ
Academy of Athens
Athens, Greece

ISSN 0925-0042   ISSN 2214-7764 (electronic)
Solid Mechanics and Its Applications
ISBN 978-3-030-89468-9   ISBN 978-3-030-89466-5 (eBook)
https://doi.org/10.1007/978-3-030-89466-5

This Springer imprint is published by the registered company Springer Nature Switzerland AG
The registered company address is: Gewerbestrasse 11, 6330 Cham, Switzerland

This book is dedicated to my wife Maria and my children Eleftherios and Alexandra-Kalliope

Preface

Experimental mechanics is a branch of engineering mechanics. Its objective is to use experiments to characterize the mechanical behavior of materials, structures, and systems. It uses advanced techniques and technologies to study the response and failure of materials and structures with the objective of improving their performance and the efficient use of materials with minimum cost and high reliability. Displacements, strains, and stresses are measured.

Experimental mechanics is not a counterpart to theoretical and analytical methods, but rather encompasses them and goes beyond by studying the true characteristics of the problems. Theoretical methods are restricted to idealized problems, while numerical methods, like finite elements, are restricted by their underlying models. Experimental mechanics goes further by determining the strains and stresses of machines and structures in operation, which is impossible in other methods without knowledge of the nature of the applied forces, the boundary and initial conditions, and the material behavior. These factors are not known in many structures in operation like aircrafts, spacecrafts, bridges, dams, etc. Experimental data are used to validate theoretical and numerical analyses of systems manufactured from complex, inhomogeneous materials, like composites, concrete, biological materials, soil, etc. The quote: *"One good test is worth a thousand expert opinions"* by Wernher Von Braun comprises the essence of experimental mechanics.

Experimental mechanics has been helped tremendously by the advent of computers and optoelectronics which had an enormous effect on optical methods. Digital image processing took advantage of computer technology. Moiré techniques have also greatly benefited. The specimen grating could be transferred to a computer screen and be superimposed with the reference grating, a task performed previously on the specimen. Of seismic impact on experimental methods was the development of the digital image correlation (DIC) method for measurement of displacements. Points on the surface of the object under study are correlated before and after deformation with algorithms run by computers. The method can be used in structures covering the macro-, micro, and nano-scale levels. DIC has been commercialized and it is relatively easy for the novice experimentalist to setup the optical arrangement, use the software and obtain results.

The purpose of this book is to present in a clear, simple, straightforward, novel, and unified manner the most used methods of experimental mechanics of solids for the determination of displacements, strains, and stresses. Emphasis is given on the principles of operation of the various methods, not on their applications to engineering problems. For more information on the methods presented in the book, there is an extensive bibliography at the end of each chapter. The book is divided into 16 chapters.

Chapter 1 presents the principle of operation of strain gages, the potentiometer and Wheatstone bridge circuits for measuring small resistance changes, and the strain gage rosettes for the complete determination of the state of strain at a point.

Chapter 2 presents the basic background for the understanding of the optical methods of experimental mechanics. Topics covered include a historical overview, geometric optics dealing with reflection, refraction, mirrors and lenses, wave optics dealing with Huygens's principle, electromagnetic theory of light, polarization, interference, and diffraction phenomena. These topics are covered in a fundamental way easily understood by an undergraduate student. The chapter is comprehensive and no further knowledge of optics is necessary for the study of optical methods of experimental mechanics in the subsequent chapters.

Chapter 3 presents the basic principles of the geometric moiré phenomenon and its applications in experimental mechanics. Following the definition of the basic terms the moiré phenomenon, the mathematical formation of moiré fringes, the relationship between line gratings and moiré fringes, the moiré patterns formed by circular, radial and line gratings, the measurement of in-plane and out-of-plane displacements and slopes, the sharpening of moiré fringes and the moiré of moiré effect are presented.

Chapter 4 presents two high-sensitivity methods, the method of coherent or intermediate moiré and the method of moiré interferometry. Unlike geometric moiré which is based on the phenomenon of obstruction of light, the two methods are based on wave optics and use the phenomena of diffraction by a grating, two-beam interference, and Fourier transformation by a lens. Topics covered in this chapter include: diffraction by two superimposed parallel gratings and formation of moiré patterns, optical filtering, and fringe multiplication obtained by the methods of coherent moiré and moiré interferometry.

Chapter 5 presents two moiré methods that use remote gratings for the determination of the gradients of the out-of-plane displacements or the sum of the two in-plane principal stresses for opaque or transparent materials, respectively. The moiré patterns are created by projecting the rulings of one grating onto the rulings of a second grating after interaction with the specimen. The first method is based on geometric moiré and uses white light, while the second is based on the diffraction of light and uses coherent light.

Chapter 6 presents the equations of the method of caustics for reflecting surfaces to obtain the caustics created by axisymmetric ellipsoid mirrors. These equations are applied for the determination of stress intensity factors in two-dimensional opening-mode and mixed-mode crack problems.

Chapter 7 presents the basics of photoelastic stress analysis including: optical patterns obtained in the plane and circular polariscopes, the isochromatic and isoclinic fringe patterns and their properties, the compensation methods, the stress separations methods, the fringe multiplication and sharpening, the transition from model to prototype, the three-dimensional photoelasticity and the photoelastic coatings.

Chapter 8 presents the Mach-Zehnder, the Michelson, and the Fizeau-type interferometers for stress analysis. The optical patterns obtained by these interferometers are studied using the Jones calculus. At the end of the chapter, a generic interpretation of the optical patterns obtained by all interferometers is presented.

Chapter 9 presents the principles of holography and holographic interferometry and its applications to photoelastic materials. The equations of the recording and reconstruction processes using the Jones calculus and a generic explanation of both holography and holographic interferometry are presented.

Chapter 10 presents the operation of optical fibers for transfer of light and describes the interferometric and Bragg grating optical fiber sensors for strain measurement.

Chapter 11 presents the speckle effect and the methods of speckle photography, speckle interferometry, shearography, and electronic speckle pattern interferometry (ESPI) for the determination of displacements.

Chapter 12 presents the essential steps of digital image correlation (DIC) including the speckle patterning, the image digitization process, the intensity interpolation, and the correlation of images.

Chapter 13 presents the theoretical basis of the thermoelastic effect, the infrared detectors for measuring very small temperature changes, the conditions of adiabaticity, the specimen preparation, the stress separation methods, and applications of thermal stress analysis.

Chapter 14 presents the principles of contact mechanics and the major methods of indentation testing at the macro-, micro-, and nano- levels used to measure the hardness, the modulus of elasticity, and the critical stress intensity factor.

Chapter 15 presents the basic principles and advantages and disadvantages of the following six major nondestructive testing methods: Dye penetrant, magnetic particles, eddy currents, X-ray diffraction, ultrasonics, and acoustic emission.

Finally, Chap. 16 presents the hole-drilling method of measuring residual stresses. The method involves monitoring the change of strains produced when a hole is drilled in a component containing residual stresses. The relevant equations for the determination of the residual stresses from the relieved strains when a hole is drilled are established for uniaxial and biaxial residual stresses.

Today's experimentalist armed with a strong background in numerical methods and computer techniques and deep understanding of the basic experimental methods should have the tools necessary to solve problems of structural behavior and material characterization at the nano-, micro-, and macro-scale levels. Needless to say, that the most valuable virtues of an experimentalist are not his/her technical skills, but honesty and integrity. Experimental results should be presented as obtained from the tests, unbiased of any relations with analytical solutions, in case they exist. This most valuable virtue should be cultivated in classrooms and laboratories.

The book provides the basics of the most used methods of experimental mechanics for measuring displacements, strains, and stresses. It is intended to be an instrument of learning, rather than a review of published contributions. It is hoped that it will be used not only as a learning tool but also as an inspiration and basis on which the researcher, the engineer, the experimentalist, the student can develop their new own ideas to promote research in experimental mechanics of solids.

The book was written during the COVID 19 pandemic. I want to take this opportunity to thank my wife Maria and my children Eleftherios and Alexandra-Kalliope for their love and support during the writing of the book. A great word of thanks goes to Dr. Panayiotis Danoglidis of the University of Texas at Arlington for his help in the preparation of the figures of the book in a speedy and efficient way. His hard work and dedication are greatly appreciated.

Athens, Greece
2021
Emmanuel E. Gdoutos

Contents

About the Author

Emmanuel E. Gdoutos is a Full Member of the Academy of Athens, the most prestigious academic institute in Greece in the chair of Theoretical and Experimental Mechanics (2016). He received his Diploma in Civil Engineering (1971) and Ph.D. (1973) from the National Technical University of Athens (NTUA). He served as Instructor in the Chair of Theoretical and Applied Mechanics of NTUA (1974–1977), Chair Professor (1977–2015) of Applied Mechanics of the Democritus University of Thrace (DUTH), Greece, Chairman of the Department of Civil Engineering (1987–1989) of DUTH, chair of the Division of Natural Sciences of the Academy of Athens (2019), Visiting Professor at the universities of Toledo, Lehigh, Michigan Technological University, University of California at Santa Barbara and Davis, Adjunct Professor at North-western University and Clark Millikan Distinguished Visiting Professor at the California Institute of Technology.

He is a member of the European Academy of Sciences and Arts (2001), European Academy of Sciences (2008), Academia Europaea (2008), International Academy of Engineering (2010), Fellow of the American Academy of Mechanics (2007) and the New York Academy of Sciences (2001), Foreign Member of the Russian Academy of Engineering (2009), the Bulgarian Academy of Sciences (2010), the Russian Academy of Sciences (2016) and the Ukrainian Academy of Sciences (2021). He is an honorary member of the Italian Society of Fracture Mechanics (2004), the Polish (2009) and the Serbian

(2011) Societies of Theoretical and Applied Mechanics, Doctorate Honoris Causa of the University of Nis of Serbia (2013), the Russian Academy of Sciences (2010) and DUTH (2019). He is a Fellow of the American Society of Mechanical Engineers (1993), the Society for Experimental Mechanics (2004), the European Structural Integrity Society (2008), the European Society for Experimental Mechanics (2010), the International Congress on Fracture (2009) the American Association for the Advancement of Science (2012). Honorary member of the Literary Society "Parnassos", Greece (2020). He served as President of the Society of Experimental Mechanics (2013–2014), the European Structural Integrity Society (2006–2010), the European Society for Experimental Mechanics (2004–2007), and Vice-President of the International Congress on Fracture (2013–2017). He has served as President of the Hellenic Society of Linguistic Heritage (2018–now) and the Theocaris Foundation of the Academy of Athens (2018–now).

He is recipient of many awards including: "Award of Merit" (2008), "Griffith Medal" (2010) of the European Structural Integrity Society, "Theocaris Award" (2009), "Lazan Award" (2009), "Tatnall Award" (2010) "Zandman Award" (2011) of the Society of Experimental Mechanics, Medal and Diploma of the "International Academic Rating of Popularity Golden Fortune" (2009), "Paton Medal" of the Ukrainian Academy of Sciences (2009), "Jubilee Medal XV Year IAE" of the International Academy of Engineering (2009), "Award of Merit" (2010), "Theocaris Award" (2012) of the European Society for Experimental Mechanics, "Golden Sign" of the Russian Academy of Engineering (2011), SAGE Best Paper Award (2012), "Panetti-Ferrari Prize" of the Turin Academy of Sciences (2012), "Colonnetti Gold Medal" of the Italian Research Institute of Metrology (2012), "Blaise Pascal Medal in Engineering 2018," of the European Academy of Sciences, "Yokobori Medal" of the International Congress of Fracture (2017), "Archon, Teacher of the Nation" of the Patriarchate of Alexandria and All Africa (2018), "Award of the Rotary Club of Mytilini" (2019), Greece, for "his invaluable contributions in engineering mechanics and higher education".

He served as President of many national and international conferences. Published more than 130 papers in international scientific journals and more than 180 in conference proceedings. Author of 9 books in Greek, 5 books in English published by international publishers (Springer, Elsevier, Kluwer Academic Publishers, Martinus Nijhoff Publishers), editor of 24 books published by the above publishers. His book *Fracture Mechanics-An Introduction, 3rd ed* by Springer accompanied by "Solutions Manual" is used as a textbook by many universities worldwide. Editor-in-chief of *Strain-An International Journal of Experimental Mechanics*, (2007–2010), guest editor of special issues of international journals, associate editor, and member of the editorial board of international scientific journals. Editor of the series of "Springerbriefs in Structural Mechanics" of Springer. He presented more many plenary/invited lectures in conferences/universities. His research interest includes problems of the theory of elasticity, use of complex functions for the solution of singularity problems of mechanics, fracture mechanics, experimental mechanics (with emphasis on the optical methods), mechanics of composite materials, sandwich structures, and nanotechnology (composite nanomaterials). In 2015 the journal *Meccanica* devoted a special issue in his honor. He is an honorary citizen of the municipality of Lesvos (Greece).

Emmanuel is married to Prof. Maria Konsta-Gdoutos. They have two children, a son Eleftherios, 34, who graduated from Northwestern University with BS Diplomas in Mechanical Engineering and Computer Science and received a Doctorate Degree in Aeronautics from Caltech, and a daughter Alexandra-Kalliope, 30, who graduated from Northwestern University with a BS Diploma in Civil Engineering. Eleftherios is currently a senior research Fellow at Caltech and Alexandra-Kalliope has a consultant appointment in a major company in the USA.

Chapter 1
Electrical Resistance Strain Gages

1.1 Introduction

An electrical resistance strain gage is a resistor used to measure strain. In its basic form, it is a wire adhered onto the surface of a component so that strains of the component are transmitted to the wire. Its operation is based on the discovery made by Lord Kelvin in 1856 that the electrical resistance of a wire increases with increasing strain and decreases with decreasing strain. By measuring the change of the resistance of the strain gage the strain is inferred. Strain gages constitute the most widely used method of measuring strain at a point.

In the following, we present the principles of operation of strain gages, the potentiometer and Wheatstone bridge circuits for measuring small resistance changes, and the strain gage rosettes for the complete determination of the state of strain at a point.

1.2 Basic Principles

We will develop a relation between the change of resistance of a wire and the strain applied to the wire. From this relation, we will be able to determine the strain by measuring the change of resistance.

Consider a conductor of uniform cross section A (in mm^2) and length L (in m) with specific resistance ρ (defined for 1 m length, 1 mm^2 cross section at 20 °C). The electrical resistance R (in Ohms, Ω) of the conductor is given by

$$R = \rho \frac{L}{A} \tag{1.1}$$

Let the resistance of the conductor change by dR, the specific resistance by $d\rho$ and the cross section area by dA when the length of the wire changes by dL. Differentiating Eq. (1.1) we obtain

© The Author(s), under exclusive license to Springer Nature Switzerland AG 2022
E. E. Gdoutos, *Experimental Mechanics*, Solid Mechanics and Its Applications 269,
https://doi.org/10.1007/978-3-030-89466-5_1

$$dR = \frac{L}{A}d\rho + \frac{\rho}{A}dL - \frac{\rho L dA}{A^2} \tag{1.2}$$

From Eqs. (1.1) and (1.2) we obtain

$$\frac{dR}{R} = \frac{d\rho}{\rho} + \frac{dL}{L} - \frac{dA}{A} \tag{1.3}$$

Let calculate the change of the cross section area dA. If V is the volume of the conductor we have

$$V = A L \tag{1.4}$$

The change of volume dV when the conductor is stretched is

$$dV = A\, dL + L\, dA = V_f - V = L_f A_t - A L \tag{1.5}$$

where L, A, and V are the length, area, and volume before stretching, and L_f, A_f, and V_f are the corresponding quantities after stretching of the conductor.

If ε is the applied uniaxial strain in the conductor, after omitting infinitesimal quantities of second order of strain ε (which are negligible for small strains), Eq. (1.5) becomes

$$dV = L(1 + \varepsilon)A(1 - v\varepsilon)^2 - AL = A(1 - 2v)dL \tag{1.6}$$

where v is the Poisson's ratio.

From Eqs. (1.5) and (1.6) we obtain

$$\frac{dA}{A} = -2v\frac{dL}{L} \tag{1.7}$$

Introducing the value of dA/A from Eq. (1.7) into Eq. (1.3) we have

$$\frac{dR}{R} = \frac{d\rho}{\rho} + (1 + 2v)\frac{dL}{L} \tag{1.8}$$

Introducing in Eq. (1.8) the axial strain $\varepsilon = dL/L$ we obtain

$$\frac{dR/R}{\varepsilon} = (1 + 2v) + \frac{d\rho/\rho}{\varepsilon} \tag{1.9}$$

The right hand side of the above equation consists of two parts: the first part $(1 + 2v)$ reflects the geometric changes in the conductor and the second term is the changes in the electrical conductivity due to the applied strain ε. Observe that strain ε appears on both sides of Eq. (1.9). In order to obtain strains from resistance measurements calibration is required. From experiments performed on certain alloys by measuring the change in resistance in terms of the applied strain ε it was concluded that the change of specific resistance $(d\rho)$ is proportional to strain ε. Thus, the second term of Eq. (1.9) is constant.

We introduce now the **sensitivity factor S_A** of the strain gage by

$$S_A = \frac{dR/R}{\varepsilon} \tag{1.10}$$

which is defined as the resistance change per initial resistance divided by the applied strain. S_A characterizes the transfer of strain to change of resistance, which can be measured electrically. Since S_A is constant for certain alloys, Eq. (1.10) allows the determination of strain ε by measuring the change of resistance dR.

We will derive an expression of the sensitivity factor S_A for large strains. We obtain from Eqs. (1.1) and (1.4) for the resistance R of a conductor in terms of its volume

$$R = \rho \frac{L^2}{V} \tag{1.11}$$

By differentiating Eq. (1.11) and following the same procedure as above we obtain

$$\frac{dR}{R} = \frac{d\rho}{\rho} + 2\frac{dL}{L} - \frac{dV}{V} \tag{1.12}$$

For plastic deformation $d\rho/\rho \to 0$ and $dV = 0$. Then we obtain from Eq. (1.12) for the sensitivity factor S_A

$$S_A = \frac{dR/R}{dL/L} = \frac{\Delta R/R}{\varepsilon} = 2 \tag{1.13}$$

Equation (1.13) indicates that the conductor sensitivity factor for small plastic deformations is equal to two. For large plastic strains, we obtain from Eq. (1.13)

$$\int_{R_0}^{R} \frac{dR}{R} = 2 \int_{L_0}^{L} \frac{dL}{L} \tag{1.14}$$

where R_0 and R refer to the initial and final resistance and L_0 and L to the initial and final length of the conductor.

or

$$\ln \frac{R}{R_0} = 2\ln \frac{L}{L_0} \quad \text{or} \quad \ln\left(1 + \frac{\Delta R}{R_0}\right) = \ln\left(1 + \frac{\Delta L}{L_0}\right)^2 \tag{1.15}$$

Equation (1.15) renders

$$\frac{\Delta R}{R_0} = 2\frac{\Delta L}{L_0} + \left(\frac{\Delta L}{L_0}\right)^2 \tag{1.16}$$

or

$$S_A = \frac{dR/R_0}{dL/L_0} = 2 + \varepsilon, \quad \varepsilon = \Delta L/L_0 \text{ (engineering strain)} \tag{1.17}$$

Equation (1.17) indicates that the strain sensitivity S_A for large plastic strains is not constant, but varies with engineering strain.

Table 1.1 presents the sensitivity factor S_A for certain alloys used in strain gage manufacturing. Note that for four alloys the value of S_A is close to 2, while for two alloys it takes the values 3.6 and 4. Referring to Eq. (1.9) note that its first term (1 + 2ν) takes values between 1.6 and 1.7 (since Poisson's ratio ν in most metals is between 0.3 and 0.35). For the alloys with S_A close to two, the second term varies between 0.3 and 0.4, and the effect of change of specific resistance on S_A is small (about 20% only). However, for alloys with S_A equal to 3.6 and 4, the contribution of the change of specific resistance is large.

An alloy used for most strain gages is the Advance or Constantan. The variation of $\Delta R/R$ versus engineering strain ε for this alloy is shown in Fig. 1.1. Note that the curve is linear and its slope which represents the strain sensitivity $S_A = 2$ is constant for

Table 1.1 Strain sensitivity S_A for certain alloys used in strain-gage manufacturing

Platinum-Tungsten (Pt 92%, W 8%)	4.0
Isoelastic (Fe 55.5%, Ni 36%, Cr 8%, Mn 0.5%)	3.6
Constantan or Advance (Ni 45%, Cu 55%)	2.1
Nichrome V (Ni 80%, Cr 20%)	2.1
Karma (Ni 74%, Cr 20%, Al 3%, Fe 3%)	2.0
Armour D (Fe 70%, Cr 20%, Al 10%)	2.0

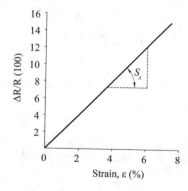

Fig. 1.1 Change of $\Delta R/R$ versus strain ε for Advance alloy. The slope of the line represents the sensitivity factor S_A

strains up to 8%. The sensitivity does not change significantly as the alloy is deformed from elastic to plastic deformation. This alloy has a high specific resistance $\rho = 0.49$ $\mu\Omega$ m and presents excellent thermal stability. Karma and Isoelastic alloys are also widely used in strain gage manufacturing. The Nichrome V, Platinum-Tungsten, and Armor D alloys are used in very specific applications.

1.3 Bonded Resistance Strain Gages

In principle, a single small length of wire bonded on the surface of the component under investigation can serve as a gage to measure the strain at a point. Circuit requirements set a lower limit of approximately 100 Ω on the gage resistance. A gage made of the finest wire with such resistance is about 100 mm long. In order to reduce the length of the gage the wire is formed into grid type and is bonded to the specimen with adhesives.

Today, metal foil electrical resistance strain gage sensors are most often used in applications. A typical foil strain gage is shown schematically in Fig. 1.2. The conductor with meander shape is printed or etched on the gage carrier material. The strain sensitive pattern is oriented along the direction of stain measuring. The strain gage consists of a sensing element attached to a thin film. The purpose of the film is

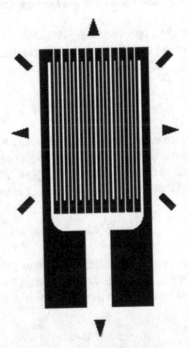

Fig. 1.2 A typical foil strain gage (courtesy of Micro-Measurements Inc.)

to serve as an insulator and carrier of the sensing element. The strain gage is bonded on the surface under consideration by an adhesive. The strain to be measured is transferred from the deformed material to the strain gage through the adhesive.

1.4   Transverse Sensitivity and Gage Factor

The strain sensitivity of a strain gage along its axis S_A was defined by Eq. (1.10). Generally, the state of strain at the point of interest is not uniaxial along the axis of the gage but biaxial. It is dictated by the values of the normal strains along the axis of the gage and the transverse direction and the shear strain. The sensing gage is sensitive not only to the axial strain along the axial segments of the strain gage grid pattern but also to the other two strains, the transverse and the shear. These strains are transferred to the sensing gage through the adhesive and the gage carrier material. The response of the gage to the shear strain is small and it can be omitted. Transverse strain sensitivity refers to the response of the gage to strains that are perpendicular to the axis of the gage. The ideal situation is that of strain gages that are completely insensitive to transverse strain. This is the case of plane wire strain gages. The transverse sensitivity of these gages is because a portion of the wire in the end loop lies in the transverse direction. This is not the case for foil strain gages whose transverse sensitivity arises from the grid design and gage construction in a complex manner.

A strain gage has two gage factors as determined in a uniaxial strain field with the gage axes aligned parallel and perpendicular to the strain field. The change of the resistance of the gage, ΔR, per initial resistance, R, may be written as

$$\frac{\Delta R}{R} = S_a \varepsilon_a + S_t \varepsilon_t \tag{1.18}$$

where S_a is the sensitivity factor of the gage to axial strain, S_t is the sensitivity factor to transverse strain, ε_a is the normal strain along the axial direction and ε_t is the normal strain along the transverse direction of the gage. The sensitivity factors S_a and S_t are dimensionless. Equation (1.18) can be put in the form

$$\frac{\Delta R}{R} = S_a (\varepsilon_a + K_t \varepsilon_t) \tag{1.19}$$

where $K_t = S_t / S_a$ is the **transverse sensitivity factor** of the gage.

The manufacturers of strain gages provide a constant known as **gage factor** S_g which relates the change of resistance to the axial strain as

$$\frac{\Delta R}{R} = S_g \varepsilon_a \tag{1.20}$$

The gage factor S_g is dimensionless.

Note that the gage factor S_g directly relates the axial strain to the change of resistance. S_g is determined by calibration experiments.

A calibration experiment often used is that of a beam in pure bending. It this case

$$\varepsilon_t = -\nu_0 \varepsilon_a \tag{1.21}$$

where ν_0 is the Poisson's ratio of the beam material.

From Eqs. (1.19) and (1.21) we obtain

$$\frac{\Delta R}{R} = S_a \varepsilon_a (1 - \nu_0 K_t) \tag{1.22}$$

Then Eqs. (1.20) and (1.22) render for the gage factor

$$S_g = S_a (1 - \nu_0 K_t) \tag{1.23}$$

We will now determine the error made when Eq. (1.20) is employed for the determination of strains. No error occurs for a uniaxial stress field or for a zero transverse sensitivity of the gage. We have from Eqs. (1.19) and (1.23)

$$\varepsilon_a = \frac{\Delta R / R}{S_g} \frac{1 - \nu_0 K_t}{1 + K_t \left(\frac{\varepsilon_t}{\varepsilon_a}\right)} \tag{1.24}$$

The apparent strain ε_a' using the gage factor is

$$\varepsilon_a' = \frac{\Delta R / R}{S_g} \tag{1.25}$$

From Eqs. (1.24) and (1.25) we have

$$\varepsilon_a = \varepsilon_a' \frac{1 - \nu_0 K_t}{1 + K_t \left(\frac{\varepsilon_t}{\varepsilon_a}\right)} \tag{1.26}$$

The percentage error $\mathcal{E}$ in neglecting the transverse sensitivity of the gage is

$$\mathcal{E} = \frac{\varepsilon_a' - \varepsilon_a}{\varepsilon_a}(100) = \frac{K_t \left(\frac{\varepsilon_t}{\varepsilon_a} + \nu_0\right)}{1 - \nu_0 K_t}(100) \tag{1.27}$$

Equation (1.27) indicates that the error $\mathcal{E}$ depends on K_t and the strain biaxiality ratio $\varepsilon_t / \varepsilon_a$. Figure 1.3 presents the variation of the error $\mathcal{E}$ versus K_t for various values of $\varepsilon_t / \varepsilon_a$ according to Eq. (1.27). Note that the error becomes significant for large values of K_t and $\varepsilon_t / \varepsilon_a$.

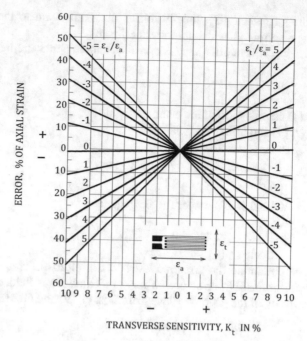

Fig. 1.3 Error, $\mathcal{E}$, versus transverse sensitivity factor, K_t, for various values of the biaxial strain ratio, $\varepsilon_t/\varepsilon_a$

1.5 Electrical Circuits

1.5.1 Introduction

Equation (1.20) indicates that the strain, ε_a, is proportional to the change of resistance of the strain gage, ΔR. In order to determine the strain, the change in the resistance needs to be measured. Strain gage resistance changes are very small, approximately 240×10^{-6} Ω per microstrain for a 120 Ω strain gage. A strain of 1000 μm/m will produce a resistance change of 0.24 Ω. Special electric circuits are used to measure such small resistance changes in terms of voltage output changes. In the following, we will present two most commonly used electrical circuits, the potentiometer circuit and the Wheatstone bridge circuit for measuring small resistance changes.

1.5.2 The Potentiometer Circuit

The potentiometer circuit is shown in Fig. 1.4. It consists of a constant voltage supply, E_i, the ballast resistor R_b and the strain gage resistor R_g. The output voltage E_T is

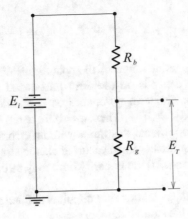

Fig. 1.4 Potentiometer circuit

obtained as

$$E_T = \frac{R_g}{R_g + R_b} E_i = \frac{1}{1 + r} E_i, \qquad (1.28)$$

where $r = R_b/R_g$.

Let us now consider that the resistances R_b and R_g change by ΔR_b and ΔR_g, respectively, and the output voltage changes by ΔE_T. Then, we obtain from Eq. (1.28)

$$E_T + \Delta E_T = \frac{(R_g + \Delta R_g)}{(R_g + \Delta R_g) + (R_b + \Delta R_b)} E_i \qquad (1.29)$$

Equations (1.28) and (1.29) render

$$\Delta E_T = \frac{r}{(1 + r)^2} \left(\frac{\Delta R_g}{R_g} - \frac{\Delta R_b}{R_b} \right)(1 - \eta) E_i \qquad (1.30)$$

where

$$\eta = 1 - \left[1 + \frac{1}{1 + r} \left(\frac{\Delta R_g}{R_g} - r \frac{\Delta R_b}{R_b} \right) \right]^{-1} \qquad (1.31)$$

With $\Delta R_b = 0$ and taking into consideration Eq. (1.20), Eq. (1.31) becomes

$$\eta = 1 - \left[1 + \frac{1}{1 + r} S_g \varepsilon \right]^{-1} \qquad (1.32)$$

Equation (1.32) can be put in the form

$$\eta = x - x^2 + x^3 - x^4 + \ldots, \quad |x| = \left| \frac{S_g \varepsilon}{1 + r} \right| < 1 \tag{1.33}$$

We will now investigate the range of the potentiometer circuit. Equation (1.30) indicates that the range depends on the error introduced in the nonlinear term η expressed by Eq. (1.33). The term η depends on r, S_g and ε. Let us consider the case of $r = 9$, $S_g = 2$ and $\varepsilon = 10\%$. Then, $x = 0.02$ and $\eta \approx 0.02$. Equation (1.30) indicates that ΔE_T is proportional to strain ε with an error of 2%. The range of the potentiometer circuit is sufficient for measuring elastic strain in metallic materials, however, for large plastic strains the linearity limit may be exceeded and corrections are required.

The **sensitivity** S_c of the potentiometer circuit is defined as the ratio of the output voltage ΔE_T divided by the strain ε as

$$S_c = \frac{\Delta E_T}{\varepsilon} \tag{1.34}$$

S_c is measured in Volt (V).
Using Eq. (1.30) with $\Delta R_b = 0$ we obtain

$$S_c = \frac{r}{(1 + r)^2} S_g E_i \tag{1.35}$$

We introduce now the power P_g dissipated by the strain gage from

$$P_g = I^2 R_g \tag{1.36}$$

where I is the current passing through the potentiometer circuit.

$$\text{Since } E_j = I(1 + r) R_g \tag{1.37}$$

we obtain

$$S_c = \frac{r}{1 + r} S_g \sqrt{P_g R_g} \tag{1.38}$$

Equation (1.38) gives the sensitivity S_c of the potentiometer circuit in terms of the ratio $r = R_b/R_g$, the power P_g dissipated by the strain gage and the strain gage resistance R_g. The first quantity $r/(1 + r)$ is related to the potentiometer circuit. It takes its maximum value of one when the ballast resistance R_b is large. However, for large values of R_b the voltage E_i required to drive the system becomes large. The second quantity $S_g \sqrt{P_g R_g}$ depends on the strain gage. Proper selection of this quantity increases the sensitivity.

Temperature compensation in the potentiometer circuit can be achieved by replacing the ballast resistor with a strain gage similar to that used in the component. The strain gage of the circuit is called *active strain gage*, while the strain gage replacing the ballast resistor is called *dummy strain gage*. The change of the resistance of the active gage is due to the strain and temperature. The change of the resistance of the dummy gage is due to change of temperature only. Equation (1.30) indicates that the temperature terms of the active and dummy gages are canceled out and, therefore, the sensitivity of the potentiometer depends only on the strain of the active gage. The dummy gage should be mounted on a material identical to that of the component under study and be exposed to the same thermal environment as that of the active gage.

The potentiometer circuit is employed to measure the resistance changes of strain gages due to dynamic loads. Usually, it is not employed for static loads. This is because in dynamic loads a filter can be used to block out the dc current E_T while allowing the passing of the voltage pulse ΔE_T. On the contrary, in static loads, the voltage E_T which appears at the output terminals eliminates the small voltage change ΔE_T due to the change of the resistance of the strain gage.

1.5.3 The Wheatstone Bridge

The Wheatstone bridge is an electrical circuit that is employed to measure an unknown electrical resistance. In our case, it is used to measure the change of the resistance a strain gage undergoes when it is subjected to strain. It consists of a constant voltage source, E_i, four resistors R_1, R_2, R_3, and R_4 arranged in a bridge configuration (Fig. 1.5), and a readout circuit.

We will relate the output voltage E_0 to the input voltage of the bridge E_i using Kirchhoff's circuit laws. Consider the bridge $ABCD$ (Fig. 1.5). The voltage drop V_{AB} across R_1 is

$$V_{AB} = \frac{R_1}{R_1 + R_2} E_i \tag{1.39}$$

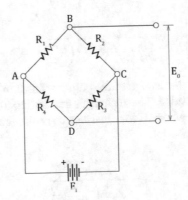

Fig. 1.5 Wheatstone bridge circuit

Similarly, the voltage drop V_{AD} across R_4 is

$$V_{AD} = \frac{R_4}{R_3 + R_4} E_i \tag{1.40}$$

The output voltage E_0 from the bridge is

$$E_0 = V_{BD} = V_{AB} - V_{AD} = \frac{R_1 R_3 - R_2 R_4}{(R_1 + R_2)(R_3 + R_4)} E_i \tag{1.41}$$

Equation (1.41) gives the output voltage in terms of the input voltage and the four resistors of the bridge. The initial output voltage E_0 from the bridge vanishes ($E_0 = 0$) and the bridge is considered **in balance** when

$$R_1 R_3 = R_2 R_4 \tag{1.42}$$

Let us now consider an initially balanced bridge and change the values of the resistances R_1, R_2, R_3, and R_4 by ΔR_1, ΔR_2, ΔR_3, and ΔR_4, respectively. Such changes in resistance can be due to changes in strain or temperature of the resistors of the bridge. The voltage output from the bridge is obtained from Eq. (1.41) as

$$\Delta E_o = \frac{(R_1 + \Delta R_1)(R_3 + \Delta R_3) - (R_2 + \Delta R_2)(R_4 + \Delta R_4)}{(R_1 + \Delta R_1 + R_2 + \Delta R_2)(R_3 + \Delta R_3 + R_4 + \Delta R_4)} E_i \tag{1.43}$$

From Eq. (1.43) with $R_1 R_3 = R_2 R_4$ and $R_2/R_1 = r$, and neglecting second order terms we obtain

$$\Delta E_0 = \frac{r}{(1+r)^2} \left(\frac{\Delta R_1}{R_1} - \frac{\Delta R_2}{R_2} + \frac{\Delta R_3}{R_3} - \frac{\Delta R_4}{R_4} \right)(1 - \eta) E_i \tag{1.44}$$

where the error term η is given by

$$\eta = \left[1 + \frac{1 + r}{\frac{\Delta R_1}{R_1} + \frac{\Delta R_4}{R_4} + r \left(\frac{\Delta R_2}{R_2} + \frac{\Delta R_3}{R_3} \right)} \right]^{-1} \tag{1.45}$$

Equation (1.44) gives the change of the output voltage in terms of the changes of the four resistors of the bridge. These changes can be due to strain or temperature.

For the common case when all four resistances of the bridge are equal ($R_1 = R_2 = R_3 = R_4$) we have

$$\eta = \frac{\sum_{i=1}^{4} \frac{\Delta R_i}{R_i}}{\sum_{i=1}^{4} \frac{\Delta R_i}{R_i} + 2} \tag{1.46}$$

From Eq. (1.46) it is deduced that for $\Delta R_1 = -\Delta R_4$ and $\Delta R_2 = \Delta R_3 = 0$ or for $\Delta R_2 = -\Delta R_3$ and $\Delta R_1 = \Delta R_4 = 0$ the error term η is zero. Also, η is zero when four active gages of the same nominal resistance are used in the bridge. For a single active gage, the error term η is 1% when $\Delta R_1/R_1 = 0.02$ (a value that corresponds to a strain of 10,000 μm/m).

The **sensitivity** S_c of the Wheatstone bridge is defined as

$$S_c = \frac{\Delta E_o}{\varepsilon} \tag{1.47}$$

S_c is measured in Volts (V).

By taking into account Eq. (1.44) with $\eta = 0$ obtain

$$S_c = \frac{r}{(1+r)^2}\left(\frac{\Delta R_1}{R_1} - \frac{\Delta R_2}{R_2} + \frac{\Delta R_3}{R_3} - \frac{\Delta R_4}{R_4}\right)\frac{E_i}{\varepsilon} \tag{1.48}$$

Considering a multiple-gage circuit with n gages ($n = 1, 2, 3, 4$) whose outputs sum when placed in the bridge circuit, we can write using Eq. (1.20)

$$\sum_{m=1}^{m=n} \frac{\Delta R_m}{R_m} = n\frac{\Delta R}{R} = nS_g\varepsilon \tag{1.49}$$

and Eq. (1.48) becomes

$$S_c = \frac{r}{(1+r)^2}nS_gE_i \tag{1.50}$$

Equation (1.50) gives the sensitivity S_c of the bridge in terms of the number n of the active arms, the gage factor S_g, the input voltage E_i and the ratio of the $r = R_2/R_1$. The sensitivity S_c becomes maximum for $r = 1$ (because the factor $r/(1 + r)^2$ becomes maximum and equal to 0.25 for $r = 1$). For $n = 4$ (four active arms) and $r = 1$ a maximum sensitivity of $S_c = S_gE_i$ is obtained, whereas with one active gage ($n = 1$) and $r = 1$, $S_c = (S_gE_i)/4$.

We will now consider the following most usual cases of strain gage arrangements in the Wheatstone bridge circuit:

Case 1. A single active gage is placed in position of resistance $R_1 = R_g$.

The power P_g dissipated by the gage is determined by

$$E_i = I_g(R_1 + R_2) = (1 + r)I_gR_g = (1 + r)\sqrt{P_gR_g} \tag{1.51}$$

Equation (1.50) with $n = 1$ taking into account Eq. (1.51) becomes

$$S_c = \frac{r}{1+r}S_g\sqrt{P_gR_g} \tag{1.52}$$

Equation (1.52) indicates that the circuit sensitivity of the bridge is due to two factors: the factor $r/(1 + r)$ which represents the circuit efficiency and the term $S_g \sqrt{P_g R_g}$ which depends on the gage. The first factor increases and takes its maximum value of *one* for large values of the resistance R_2 (when r becomes large the factor $r/(1 + r)$ tends to 1). However, as the value of r increases the voltage supply E_i increases. No temperature compensation for this circuit can be achieved. For $r = 9$, $S_c = 0.9 S_g \sqrt{P_g R_g}$.

Case 2. A single active gage is placed in position of resistance $R_1 = R_g$ and a dummy gage in position of resistance $R_2 = R_g$. Let I_g be the current that passes through both gages. We have

$$E_i = 2 I_g R_g = 2\sqrt{P_g R_g} \tag{1.53}$$

Equation (1.50) with $r = 1$ and $n = 1$ gives

$$S_c = \frac{1}{2} S_g \sqrt{P_g R_g} \tag{1.54}$$

The position of a dummy gage in this arrangement ensures temperature compensation. The sensitivity of this arrangement is reduced to 50% of the sensitivity of the previous case (for large values of the resistance R_2). Note that S_c does not depend on r as in the previous case but only on the gage selection.

Case 3. A single active gage is placed in position of resistance $R_1 = R_g$ and a dummy gage in position of resistance $R_4 = R_g$. Fixed resistors are places in positions R_2 and R_3.

We have

$$E_i = I_g(R_1 + R_2) = (1 + r)I_g R_g = (1 + r)\sqrt{P_g R_g} \tag{1.55}$$

where I_g is the current that passes through both gages.

Substituting these values into Eq. (1.50) with $n = 1$ we obtain for the circuit sensitivity

$$S_c = \frac{r}{1 + r} S_g \sqrt{P_g R_g} \tag{1.56}$$

The sensitivity is the same as in case 1. The circuit is temperature compensated. Thus, by placing a dummy gage in position R_4 temperature compensation is achieved without any loss in circuit sensitivity.

Case 4. Four active gages are placed in the bridge, with one gage in each arm of the bridge ($R_1 = R_2 = R_3 = R_4 = R_g$).

We have

$$E_i = 2 I_g R_g = 2\sqrt{P_g R_g} \tag{1.57}$$

When the signals of the four gages add (for example when the gages R_1 and R_3 are placed on one side of a cantilever beam under pure bending and the gages R_2 and R_4 are placed on the opposite side of the beam) we obtain from Eq. (1.50) with $n = 4, r = 1$

$$S_c = 2S_g\sqrt{P_g R_g} \tag{1.58}$$

The bridge with four active arms is slightly more than two times sensitive than the single-active-arm bridges of cases 1 and 3. This bridge is temperature compensated. The increased sensitivity is paid by the higher cost of using four gages.

By comparing the above four bridge arrangements we conclude that the sensitivity of the circuit varies between $0.5\sqrt{P_g R_g}$ and $2\sqrt{P_g R_g}$. Usually, for strain measurements in experimental mechanics single-active-arm bridges are employed to avoid the cost of installing additional gages, and the signal obtained from the bridge is significantly amplified.

1.6 Strain Gage Rosettes

The state of strain at a point is defined by the three strain components ε_{xx}, ε_{yy}, and γ_{xy} in a Cartesian system O_{xy}. The principal strains ε_1 and ε_2 and the principal angle β are determined by

$$\varepsilon_1 = \frac{1}{2}\left(\varepsilon_{xx} + \varepsilon_{yy}\right) + \frac{1}{2}\sqrt{\left(\varepsilon_{xx} - \varepsilon_{yy}\right)^2 + \gamma_{xy}^2}$$

$$\varepsilon_2 = \frac{1}{2}\left(\varepsilon_{xx} + \varepsilon_{yy}\right) - \frac{1}{2}\sqrt{\left(\varepsilon_{xx} - \varepsilon_{yy}\right)^2 + \gamma_{xy}^2}$$

$$2\beta = \tan^{-1}\frac{\gamma_{xy}}{\varepsilon_{xx} - \varepsilon_{yy}} \tag{1.59}$$

Normal strains are measured by strain gages along a certain direction. The strains ε_A, ε_B, and ε_C along the directions A, B, and C that makes angle β_A, β_B, and β_C with the x-axis (Fig. 1.6) are given by

$$\varepsilon_A = \varepsilon_{xx}\cos^2\beta_A + \varepsilon_{xx}\sin^2\beta_A + \gamma_{xy}\sin\beta_A\cos\beta_A$$

$$\varepsilon_B = \varepsilon_{xx}\cos^2\beta_B + \varepsilon_{xx}\sin^2\beta_B + \gamma_{xy}\sin\beta_B\cos\beta_B$$

$$\varepsilon_A = \varepsilon_{xx}\cos^2\beta_C + \varepsilon_{xx}\sin^2\beta_C + \gamma_{xy}\sin\beta_C\cos\beta_C \tag{1.60}$$

For the determination of the state of strain at a point measurement of three normal strains along three different directions is needed. Strain gage rosettes with three gages along prescribed directions are used for this purpose. The most common rosettes are the three-element tee (Fig. 1.7a), the **rectangular** (Fig. 1.7b), and the **delta** (Fig. 1.7c) **rosette**.

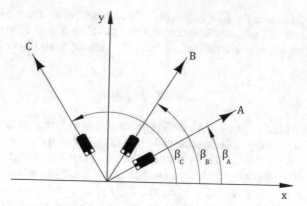

Fig. 1.6 Three strain gages at directions *A*, *B*, *C* oriented at angles β_A, β_B, β_C with respect to the x-axis

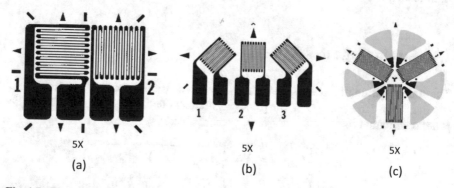

Fig. 1.7 Rectangular and delta rosettes (courtesy of Micro-Measurements Inc.)

For a rectangular rosette with three gages at angles $\beta_A = 0$, $\beta_B = 45°$, $\beta_C = 90°$ along the directions *A*, *B*, *C*, respectively, we obtain from Eq. (1.60)

$$\varepsilon_{xx} = \varepsilon_A, \, \varepsilon_{yy} = \varepsilon_C, \, \gamma_{xy} = 2\varepsilon_B - \varepsilon_A - \varepsilon_c \tag{1.61}$$

The principal strains ε_1 and ε_2 and the principal angle β are given by

$$\varepsilon_1 = \frac{1}{2}(\varepsilon_A + \varepsilon_C) + \frac{1}{2}\sqrt{(\varepsilon_A - \varepsilon_C)^2 + (2\varepsilon_B - \varepsilon_A - \varepsilon_C)^2}$$

$$\varepsilon_2 = \frac{1}{2}(\varepsilon_A + \varepsilon_C) - \frac{1}{2}\sqrt{(\varepsilon_A - \varepsilon_C)^2 + (2\varepsilon_B - \varepsilon_A - \varepsilon_C)^2}$$

$$2\beta = \tan^{-1}\frac{2\varepsilon_B - \varepsilon_A - \varepsilon_C}{\varepsilon_A - \varepsilon_C} \tag{1.62}$$

For the delta rosette with three gages at angles $\beta_A = 0$, $\beta_B = 60°$, $\beta_C = 120°$ along the directions A, B, C, respectively, the principal strains ε_1 and ε_2 and the principal angle β are given by

$$\varepsilon_1 = \frac{1}{3}(\varepsilon_A + \varepsilon_B + \varepsilon_C) + \frac{1}{2}\sqrt{\left(\varepsilon_A - \frac{\varepsilon_A + \varepsilon_B + \varepsilon_C}{3}\right)^2 + \left(\frac{\varepsilon_B - \varepsilon_C}{\sqrt{3}}\right)^2}$$

$$\varepsilon_2 = \frac{1}{3}(\varepsilon_A + \varepsilon_B + \varepsilon_C) - \frac{1}{2}\sqrt{\left(\varepsilon_A - \frac{\varepsilon_A + \varepsilon_B + \varepsilon_C}{3}\right)^2 + \left(\frac{\varepsilon_B - \varepsilon_C}{\sqrt{3}}\right)^2}$$

$$2\beta = \tan^{-1}\left(\frac{1}{\sqrt{3}} \frac{(\varepsilon_A - \varepsilon_C)}{\varepsilon_A - \frac{\varepsilon_A + \varepsilon_B + \varepsilon_C}{3}}\right) \tag{1.63}$$

Further Readings

1. Perry CC, Lissner HR (1962) Strain gage primer. McGraw-Hill
2. Window AL, Holister GS (1982) Strain gauge technology. Applied Science
3. Hoffmann K (1989) An introduction to measurements using strain gages. Hottinger Baldwin Messtechnik
4. Dally JW, Riley WF (1991) Experimental stress analysis. McGraw-Hill, pp 366–374, 129–261
5. Hannah RL, Reed SE (eds) (1992) Strain gage users' handbook. Kluwer
6. Murray WM, Miller WR (1992) The bonded electrical resistance strain gage: An introduction. Oxford University Press
7. Window AL (ed) (1992) Strain gauge technology, 2nd ed. Springer
8. Dally JW, Riley WF, Sirkis JS (1993) Strain gages. In: Kobayashi AS (ed) Handbook of experimental mechanics, 2nd ed. Society for Experimental Mechanics, pp 39–77
9. McConnell KG, Riley WF (1993) Strain-gage instrumentation and data analysis. In: Kobayashi AS (ed) Handbook of experimental mechanics, 2nd ed., Society for Experimental Mechanics, pp 79–117
10. Khan AS, Wang X (2000) Strain measurements and stress analysis. Pearson
11. Kallas N, Sathianathan D, Wu X (2005) Designing a weighing system using strain gages, 3rd ed. Hayden-McNeil Publishing
12. Watson RB (2008) Bonded electrical resistance strain gages. In: Sharpe WN (ed) Handbook of experimental solid mechanics. Springer, pp 283–333
13. Khan AS, Wang X (2001) Strain measurements and stress analysis. Prentice Hall, pp 30–93
14. Holman JP (2012) Experimental methods for engineers, 8th ed. McGraw-Hill, pp 505–516
15. Sciammarella CA, Sciammarella FM (2012) Experimental mechanics of solids. Wiley, pp 41–121
16. Shukla A, Dally JW (2014) Experimental solid mechanics, 2nd ed. College House Enterprises, pp 149–256
17. Keil S (2017) Technology and practical use of strain gages with particular consideration of strain analysis using strain gages. Wiley
18. Micro Measurements (2013) Noise control in strain gage measurements, Tech Note 501
19. Micro Measurements (2010) Optimizing strain gage excitation levels, Tech Note 502
20. Micro Measurements (2014) Strain gage thermal output and gage factor variation with temperature, Tech Note 504
21. Micro Measurements (2018) Strain gage selection: criteria, procedures, recommendations, Tech Note 505

22. Micro Measurements (2010) Bondable resistance temperature sensors and associated circuitry, Tech Note 506
23. Micro Measurements (2010) Errors due to Wheatstone bridge nonlinearity, Tech Note 507
24. Micro Measurements (2010) Fatigue characteristics of micro-measurements strain gages, Tech Note 508
25. Micro Measurements (2018) Errors due to transverse sensitivity in strain gages, Tech Note 509
26. Micro Measurements (2010) Errors due to misalignment of strain gages, Tech Note 511
27. Micro Measurements (2010) Plane-shear measurement with strain gages, Tech Note 512
28. Micro Measurements (2013) Shunt calibration of strain gage instrumentation, Tech Note 514
29. Micro Measurements (2014) Strain gage rosettes: selection, application and data reduction, Tech Note 515
30. Micro Measurements (2010) Errors due to shared leadwires in parallel strain gage circuits, Tech Note 516

Chapter 2
Fundamentals of Optics

2.1 Introduction

Optical methods of stress analysis, including geometric, coherent, and interferometric moiré, photoelasticity, interferometry, speckle photography and interferometry, holography, optical fibers, digital image correlation, are of great importance in experimental mechanics.

This introductory chapter presents the basic background of optics for the understanding of these methods. Topics covered include: a historical overview of light theories, geometric optics dealing with reflection, refraction, mirrors and lenses, wave optics dealing with Huygens's principle, electromagnetic theory of light, polarization, interference, and diffraction phenomena. These topics are presented in a fundamental way easy to understand for an undergraduate student. The chapter is comprehensive and no further knowledge of optics is necessary for the study of the optical methods of experimental mechanics.

2.2 Historical Overview

Since the ancient Greek era up to our times' several theories of the nature of light have been proposed. These theories remained valid as long as they could explain phenomena related to the propagation of light and its interaction with matter and they were replaced by new theories when they were not able to explain newly discovered phenomena.

The Greek philosopher Aristotle and his contemporaries attempted to explain how the sensation of vision is caused, how light comes from the sun or the stars to earth, the nature of light, and the mechanism of vision. Archimedes designed mirrors capable of burning the Roman ships. Euclid studied the properties of light, postulated that it travels in straight paths, and described the laws of reflection. However, the early postulates of Greeks about the nature of light, due to the lack of experiments,

© The Author(s), under exclusive license to Springer Nature Switzerland AG 2022
E. E. Gdoutos, *Experimental Mechanics*, Solid Mechanics and Its Applications 269,
https://doi.org/10.1007/978-3-030-89466-5_2

were only of philosophical character and they were not able to formulate a systematic theory capable of explaining the optical phenomena involved in the propagation of light and its interaction with matter.

Newton in 1660 formulated the first systematic theory of light, according to which the luminous body emits streams of small particles (called corpuscles) that propagate with the velocity of light. He explained the phenomenon of reflection of light but failed to explain the phenomenon of refraction.

Newton's corpuscular theory was replaced in 1678 by the **wave theory** proposed by Huygens, according to which the luminous body emits transverse waves in a hypothetical medium filling the space, called *ether*. When these waves reach the eye produce the sensation of vision. The wave theory of light explains the phenomena of reflection, refraction, interference, and diffraction. However, the explanation of the phenomenon of polarization of light necessitates that light waves must be transverse, which, due to the great velocity of propagation of light, requires unrealistic properties of the ether. The ether would have to be considered a fluid with a great modulus of elasticity and a low density.

All the above-mentioned phenomena of interference, diffraction, and polarization of light without the need of the hypothetical medium of ether can be explained by the **electromagnetic theory** of light introduced by Maxwell in 1873. According to this theory, light waves are electromagnetic waves, which are transverse waves and propagate in space by varying electric and magnetic fields. The velocity of propagation of light is equal to the velocity of propagation of electromagnetic waves.

Phenomena concerning the interaction of light and matter, such as the photoelectric and Compton effects cannot be explained by the electromagnetic theory. These phenomena were explained by a **particle theory** of light developed by Planck and Einstein in 1905. The theory is based on the assumption of the exchange of energy between radiation and matter, such as the emission or absorption of light takes place in a discontinuous matter by definitely determined amounts of energy that depend only on the frequency of the radiation. The particle theory of light is inadequate to explain some phenomena related to the wave character of light, such as interference and diffraction.

For a unified explanation of the phenomena related to the particle and the wave nature of light, the **quantum theory** was introduced by de Broglie, Schrödinger, and Heisenberg in early 1900s. This theory attributes to light the characteristics of both particles and waves. It is a combination of the particle and the wave theories of light.

In this book, only phenomena related to geometric optics and the wave nature of light will be dealt with. No problems related to the interaction of light and matter will be studied. In this respect, for the study of the problems of the book the geometric optics and the electromagnetic and wave theories of light will be used.

2.3 Light Sources, Wave Fronts, and Rays

All matter at high temperatures emits visible light. An **incandescent light** is a bright light produced from a wire filament heated until it glows. The filament is enclosed in a bulb and current is supplied to the filament by wires. The filament usually operates at a temperature of about 3000 °C.

Coherence is a basic property of electromagnetic waves. It is a fixed relationship between the phase of waves in a light beam of a single frequency. Two monochromatic sources of the same frequency and same waveform are coherent when they have a constant phase relationship. Coherent waves are emitted from coherent light sources. Coherence enables stationary interference of waves. Two separate light sources are impossible to be coherent. To achieve coherence one light source is usually split into two or more beams by optical instruments (beam splitter mirrors).

A light source of great importance in experimental mechanics is the **laser** (acronym for "light **a**mplification by **s**timulated **e**mission of **r**adiation"). A laser, unlike an incandescent source, emits highly coherent monochromatic light in a narrow and tight path over long distances. The foundation for the laser was laid down by Albert Einstein in 1916 who introduced the concept of "stimulated emission" (a photon interacts with an excited molecule or atom and causes the emission of a second photon having the same frequency, phase, polarization, and direction). The atoms can be excited in a continuous or pulsed manner. In the first case, the energy supplied is continuous, in the second the energy is supplied by periodic pulses. The first most widely used laser in experimental mechanics is the **Helium–Neon (He–Ne) gas laser** (of about 85% He and 15% Ne, emitting red light at 632.8 nm). It was built in 1960 at Bell Labs. Lasers have many applications in medicine, in machining, in electronics. In experimental mechanics lasers are indispensable. Without laser optical methods like holography, interferometric moiré, speckle methods would have been impossible.

A beam of light is called **collimated** when it has parallel rays and diverges minimally as it propagates. Light can be made collimated by an optical device called a collimator or other processes. Lasers emit collimated light.

The concept of the **wave front** is used to describe propagation of light. A wave front is defined as the set of points at which the phase of transverse vibration of the electric field vector for a number of light rays is the same. A point light source in a homogeneous isotropic medium produces spherical wave fronts concentric with the light source. At large distances from the point light source where the radii of the spherical wave fronts are large a part of the sphere can be considered as a plane, and we have plane wave fronts. Figure 2.1 presents a plane wave front that is transformed by a lens to a spherical wave front.

It is a common experience that light propagates in straight lines. This leads to the **ray model** of light. In particle theory, light rays were used to describe light propagation long before the wave theory of light was invented. A **ray** is an imaginary line along the direction of propagation of light. It represents a narrow beam of light. Light comes from the sun in the form of straight rays. The rays are perpendicular

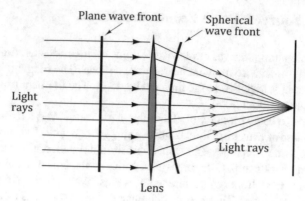

Fig. 2.1 A plane wave front transformed by a lens to a spherical wave front

to the wave fronts of light. For waves emanating from a point light source, the rays are the radii of the spherical wave fronts, while for plane wave fronts the rays are perpendicular to the fronts and parallel to each other (Fig. 2.1).

The ray model of light has been successful in describing many phenomena involved with the reflection and refraction of light and the formation of images by mirrors and lenses. It fails to describe interference and diffraction phenomena for which the wave nature of light is appropriate. The branch of optics for which the ray model of light is adequate is called **geometric optics**. The branch that deals with the wave nature of light is called **physical optics**.

In the following, we will use geometric optics to explore the phenomena of reflection and diffraction of light and the formation of images by mirrors and lenses. For the phenomenon of polarization, interference, and diffraction we will resort to the wave theory of light.

2.4   Reflection and Mirrors

2.4.1   Reflection

We will use the principles of geometric optics to explore the phenomenon of **reflection** of light. Let us define the concept of **object** as anything from which light rays radiate. An object can be self-luminous or can reflect light it receives from another source. A **point object** is an idealized concept used often in optics. When light strikes an object part of it is reflected and part is absorbed (opaque materials) or transmitted (transparent materials) by the object. For a shiny object over 95% of the light intensity, maybe reflected. Reflection from a rough surface is called **diffuse reflection**, and reflection from a mirror is called **specular reflection**.

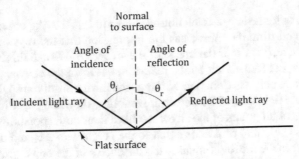

Fig. 2.2 Law of reflection. The angle of incidence θ_i is equal to the angle of reflection θ_r

When a light ray strikes a flat surface we define the **angle of incidence, θ_i,** as the angle the ray makes with the normal to the surface and the **angle of reflection, θ_r,** as the angle the reflected ray makes with the normal (Fig. 2.2). Experimental evidence indicates that:

a. The incident and reflected rays and the normal to the surface lie on the same plane.
b. The angle of reflection is equal to the angle of incidence, $\theta_r = \theta_i$.

The above two statements constitute the **law of reflection**. We will use this law to find the images of objects by plane and spherical mirrors.

2.4.2 Plane Mirrors

A **plane mirror** is a flat reflecting surface. Consider the simple case of a one-dimensional arrow object AB of height h in front of a plane mirror and let us find the image of the object from the mirror (Fig. 2.3). Each point of the object has a corresponding point of the image. Consider two rays from point B, one perpendicular to the mirror and one passing through a point O. Both rays are reflected from the

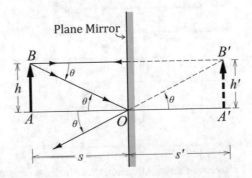

Fig. 2.3 Image formation of an object AB in front of a plane mirror

mirror according to the law of reflection as shown in Fig. 2.3. An observer who sees the rays reflected from the mirror has the impression that the rays come from the arrow $A'B'$. The arrow $A'B'$ is the image of the arrow AB. We call this image **virtual image** because it is formed not by the real rays, but from the extensions of the real rays. Real rays do not pass from point B' since the mirror is not transparent and there is no light on the right side of the mirror. If we follow the same procedure we realize that the images of all points of the arrow AB have their corresponding points on the arrow $A'B'$. The height h' of $A'B'$ is equal to the height h of AB, and the distance s' of the image from the mirror is equal to the distance s of the object from the mirror. We can see from Fig. 2.3 that the image arrow points in the same direction as the object arrow, and we say the image is **erect**.

2.4.3 Spherical Mirrors

Plane mirrors produce imaginary images of the same size as the object. When real images smaller or larger than the object are required curved mirrors are used. The most common curved mirror is **spherical,** which forms a section of a sphere. A spherical mirror is called **convex** if the incident light is reflected from the outer surface of the sphere and **concave** if the incident light is reflected from the inner surface of the sphere. In the convex mirrors, the center of curvature is on the opposite side of the reflected ray, while in concave mirrors the center of curvature is on the same side of the reflected rays.

When the spherical mirror is small compared to its radius all light rays parallel to the **principal axis** of the mirror after reflection pass through the same point which is called the **focal point F** of the mirror (Fig. 2.4). The distance between the focal point F and point A on the principal axis is called the **focal length, f,** of the mirror. Since the triangle CFB is isosceles the focal length f is half the radius r of curvature of the mirror:

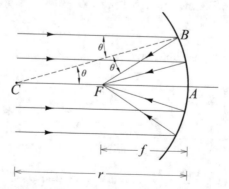

Fig. 2.4 Focal point of a concave mirror. All parallel rays incident on a spherical mirror pass through the focal point after reflection

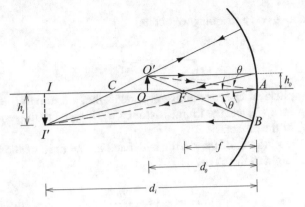

Fig. 2.5 Formation of image II' of an object OO' in front of a spherical mirror

$$f = \frac{r}{2} \tag{2.1}$$

Equation (2.1) applies for small mirrors (small angles θ).

When the angle θ is not small, not all rays parallel to the principal axis of the mirror pass through a perfect focus at F. This defect of spherical mirrors is called **spherical aberration**.

To obtain the image of a line object OO' of height h_0 at distance d_0 from the center A of a spherical mirror with focal point F and focal length f (Fig. 2.5) we apply the law of reflection for rays emanating from points of the object. To simplify the procedure we use the following three characteristic rays from point O' of the object OO': a ray parallel to the axis of the mirror which after reflection passes through the focus F, a ray that passes through the focus F which after reflection becomes parallel to the axis of the mirror and a ray that passes through the center C of curvature which is perpendicular to the mirror and is reflected back on itself (Fig. 2.5). The image II' of the object OO' is formed by the intersection of the reflected rays. It has a height h_i and is formed at a distance d_i from point A. Note that image II' is formed by real rays and is a **real image**, contrary to the plane mirror in which the image is virtual. Furthermore, note that the image is inverted relative to the object.

By applying the law of reflection for the light ray $O'A$, we obtain from the similar triangles $O'AO$ and $I'AI$ (they are orthogonal and have an angle equal)

$$\frac{h_o}{h_i} = \frac{d_o}{d_i}$$

Furthermore, we obtain from the similar triangles $O'FO$ and BFA by assuming the length of the arc AB is equal to the height of the image h_i (the arc AB is small compared to the radius of the mirror)

$$\frac{h_o}{h_i} = \frac{OF}{FA} = \frac{d_o - f}{f}$$

From the above two equations, we obtain

$$\frac{1}{d_o} + \frac{1}{d_i} = \frac{1}{f}$$

(2.2)

This is the **spherical mirror equation**. It states that the sum of the reciprocal of the **object distance, d_0,** and the **image distance, d_i,** is equal to the reciprocal of the focal length, f, of the mirror.

The **magnification**, m, of the mirror is defined as the ratio of the height of the image and the height of the object as

$$m = \frac{h_i}{h_o} = -\frac{d_i}{d_o}$$

(2.3)

In Eqs. (2.2) and (2.3) the following sign convention applies: h_o is taken positive, h_i is positive if the image is upright and negative if inverted relative to the object, d or d_i are positive if image or object is in front of the mirror. The magnification is positive for an upright image and negative for an inverted image.

Equations (2.2) and (2.3) which were derived for a concave mirror also apply for convex mirrors with the above sign conventions. However, the focal length f and the radius of curvature of convex mirrors are negative.

2.5 Refraction

When light strikes an interface between two transparent materials (such as air and glass) is partly reflected and partly passes through into the second material (Fig. 2.6a). When light makes an angle with respect to the interface the ray changes direction in the second material. The phenomenon of changing direction of light as it passes through two media is called **refraction**. Before we study the laws of refraction let us define a quantity that is very important in optics, **the index of refraction, n**.

The index of refraction of a material, n, is defined as the ratio of the speed of light in vacuum, c, to the speed of light in the material, υ:

$$n = \frac{c}{\upsilon}$$

(2.4)

The speed of light in vacuum, c, is

$$c = 2.99792458 \times 10^8 \text{ m/s}$$

which is the speed of all electromagnetic waves. It is usually rounded off to

$$c = 3 \times 10^8 \text{ m/s}$$

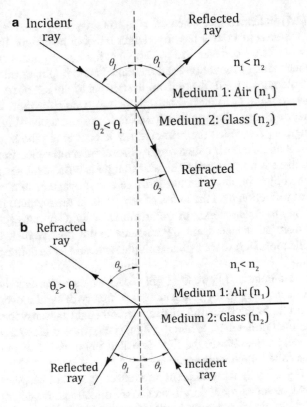

Fig. 2.6 Refraction of light at the interface between two media. Light passing from air to glass (**a**), light passing from glass to air (**b**)

The speed of light in a material, v, is less than the speed of light in vacuum ($v < c$). Thus, the index of refraction of a material is greater than $1 (n > 1)$.

The **law of refraction**, known as **Snell's law**, may be put as:

a. The incident, the reflected and the refracted rays lie in the same plane which is perpendicular to the interface between the two media.

b. The angle of refraction θ_2 into medium 2, defined as the angle subtended between the refracted ray and the normal to the surface, satisfies the following equation (Fig. 2.6a).

$$n_1 \sin \theta_1 = n_2 \sin \theta_2 \tag{2.5}$$

where n_1 is the index of refraction of medium 1, n_2 is the index of refraction of medium 2 and θ_1 is the angle of incidence.

Equation (2.5) indicated that when $n_2 > n_1$, then $\theta_2 < \theta_1$. That is, when a ray passes from a material to another having a larger index of refraction, the ray is bent toward the normal. If we consider in Fig. 2.6b that the ray originates from medium 2 (having higher index of refraction than medium 1) and impinges on the interface then the ray in medium 1 bends away from the normal to the interface.

It is evident from Eq. (2.5) that when light passes from a medium with a higher index of refraction to a medium with a lower index of refraction (Fig. 2.6b) there is a critical angle θ_1 for which the refracted ray is bent along the interface ($\theta_2 = 90°$). This critical angle $\theta_1 =$ is called **angle of total internal reflection.** For angles greater than θ_c there is no refracted light, and all light is reflected. Internal reflection occurs only when light impinges on an interface and propagates in a medium with smaller index of refraction than the index of refraction of the medium it originates.

We study now the phenomenon of polarization of light by reflection. Ordinary light, emitted from an incandescent light source is not polarized. It can become polarized by reflection. Before we explain this phenomenon let us define the meaning of polarization.

In ordinary or unpolarized light, the light vector which represents the electric or magnetic vector according to the electromagnetic theory of light moves irregularly on the plane normal to the direction of propagation (light is transverse wave) and shows no directional preference. When the light vector moves along a well-defined path (straight line, circle, ellipse) the light is called **polarized**. If it moves along a straight line it is called **linearly polarized**.

Consider in Fig. 2.7 a light ray incident on a reflecting surface between two media. The electric field vector of the light vibrates in all directions. It was found that for a particular angle of incidence, called **polarizing angle, θ_p,** the reflected light is completely (100%) linearly polarized with its plane of polarization perpendicular

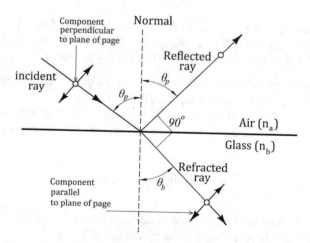

Fig. 2.7 Polarization at reflection. When light is incident on a reflecting surface at the polarizing angle the reflected light is linearly polarized (with plane of polarization perpendicular to the plane of incidence) and perpendicular to the refracted ray

to the plane of incidence (the plane that contains the incident, the reflected and the refracted rays, and the normal to the interface) and perpendicular to the refracted ray. In Fig. 2.7 the light vector along the plane of incidence is represented by arrows and the light vector perpendicular to the plane of incidence is represented by dots. This phenomenon was discovered by Sir David Brewster. The polarizing angle, θ_p, is defined by

$$\tan \theta_p = \frac{n_b}{n_a} \tag{2.6}$$

where n_a is the index of refraction of the medium of the incident and reflected rays and n_b is the index of refraction of the medium of the refracted ray.

Equation (2.6) can be derived from the electromagnetic theory of light. It is known as **Brewster's law** and the angle θ_p as **Brewster angle**. For light impinging from air on a glass surface ($n = 1.6$) the Brewster angle θ_p is obtained from Eq. (2.6) as 58° ($n_b = 1.6$, $n_a = 1$). From Eqs. (2.6) and (2.5) it is obtained that $\theta_b = 90° - \theta_p$, that is at the polarizing angle the reflected ray is perpendicular to the refracted ray.

2.6 Thin Lenses

Lenses are very important optical devices employed in optical methods of experimental mechanics. In this section, we will study lenses using the concept of light rays within the frame of geometric optics. Lenses as Fourier transform analyzers will be examined later on in this chapter in the section of diffraction.

A lens is a simple transmissive optical devise that focuses or disperses light. It is made of glass or transparent plastic, so its index of refraction is higher than that of the surrounding air. Its principle of operation is based on refraction (Snell's law). It usually consists of two spherical surfaces, even though cylindrical surfaces are possible. A **thin lens** is defined by a small diameter compared to the radii of curvature of the two lens surfaces. The two surfaces of a thin lens can be concave, convex, or flat (Fig. 2.8). The **axis** of the lens is a straight line that is perpendicular to the two surfaces of the lens and passes through its center. The lenses are divided into **converging** and **diverging**. A converging lens is thicker at the center than at the edges. A diverging lens is thinner at the center than at the edges (Fig. 2.8). When a beam of light parallel to the axis strikes a converging lens the refracted rays after passing the lens are focused to a point called the **focal point F** of the lens (Fig. 2.9a). If parallel rays fall at a lens at an angle they focus at a different point **Fa** at the same distance from the lens (Fig. 2.9). The plane of all points F and F_a are called the **focal plane** of the lens (Fig. 2.9b). When the bean strikes a diverging lens the refracted rays appear to diverge at the focal point (Fig. 2.10). A lens has two focal points, symmetrical with respect to the lens. The distance of the focal point F from the lens is called the **focal length, f**. It is the most important property of a lens.

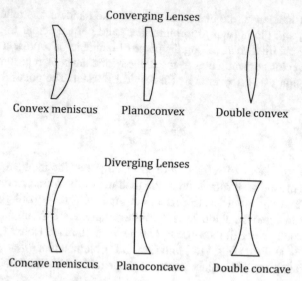

Fig. 2.8 Converging and diverging lenses

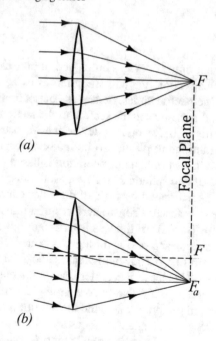

Fig. 2.9 Parallel rays are brought to a focus by a converging lens

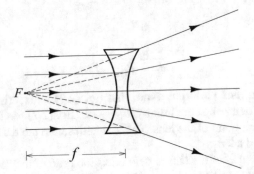

Fig. 2.10 Diverging lens. The reflected rays appear to diverge at the focal point F

To find the image of an object in front of a lens we use three characteristic rays, in a way similar to the case of curved mirrors. A ray parallel to the axis, which after refraction by the lens passes through the focal point F; a ray passing through the other focal point F' which emerges from the lens parallel to its axis; a ray passing through the center of the lens which emerges from the lens at the same angle as it entered (at the center of the lens its two surfaces are parallel). Construction of the image of an object in front of a converging lens is shown in Fig. 2.11.

Consider an object OO' of height h_0 in front of a converging lens, and find the relation between the object distance d_0, the image distance d_i, and the focal length f of the lens. From the similar triangles FII' and FAB we have

$$\frac{h_i}{h_0} = \frac{FI}{FA} = \frac{d_i - f}{f}$$

where h_i is the height of the image.

Similarly, from the similar triangles OAO' and IAI' we have

$$\frac{h_i}{h_0} = \frac{d_i}{d_0}$$

From the above two equations, we obtain

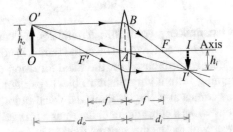

Fig. 2.11 Construction of the image II' of an object OO' in front of a converging lens

$$\frac{1}{d_o} + \frac{1}{d_i} = \frac{1}{f} \qquad (2.7)$$

This is the **thin lens equation**. Note that Eq. (2.7) is exactly the same as the mirror Eq. (2.2). For an object at infinity, $d_0 \to \infty$, Eq. (2.7) renders $d_i = f$, that is, its image is the focal point of the lens. This coincides with the definition previously of the focal point of a lens.

The magnification m of the lens is defined as the ratio of the image height to the object height. From Fig. 2.11 m is given by

$$m = \frac{h_i}{h_o} = -\frac{d_i}{d_o} \qquad (2.8)$$

The minus sign indicates that when d_0 and d_i are both positive the image is inverted, and h_0 and h_i have opposite signs.

Equations (2.7) and (2.8) are exactly the same as the corresponding Eqs. (2.2) and (2.3) of curved mirrors. The same equations apply to convex (diverging) lenses. The above equations apply with the following sign convention:

a.   The focal length f is positive for converging lenses and negative for diverging lenses.
b.   The object distance d_0 is positive if the object is on the side of the lens from which the light is coming, otherwise, it is negative.
c.   The image distance d_i is positive if the image is on the opposite side of the lens from which the light is coming, otherwise, it is negative.
d.   The height of the image, h_i, is positive if the image is upright, and negative if the image is inverted relative to the object.
e.   The height of the object h_0 is always positive.

An expression can be obtained between the focal length of a lens, f, the radii of curvature of its two surfaces R_1 and R_2, and the index of refraction n of the material of the lens. It is given by

$$\frac{1}{f} = (n - 1)\left(\frac{1}{R_1} + \frac{1}{R_2}\right) \qquad (2.9)$$

This is called the **lensmaker's equation**. R_1 and R_2 are considered positive if both surfaces are convex. For a concave surface, the radius is considered negative.

The above developments were based on the ideal situation of thin lenses. As a consequence, light coming from a very distant object (parallel beam of light) and passing through a lens comes at a single point, the focal point of the lens. This is not the case for real lenses. As a consequence, the image formed by a lens is blurred or distorted. This property of lenses that causes light to be spread out rather than to be focused to a single point is called **aberration**. Generally speaking, aberration is defined as a departure of the performance of an optical system from the predictions of the paraxial optics (small angle approximation). There are many types of aberrations in lenses, such as spherical, distortion, chromatic aberrations.

2.7 The Wave Nature of Light—Huygens' Principle

Geometric optics which is based on the concept of the light ray was successfully used in the previous sections to explain the phenomena of reflection and refraction and the formation of images by mirrors and lenses. However, geometric optics does not account for the explanation of the phenomena of interference and diffraction of light. For such phenomena, the wave or the electromagnetic theory of light is most appropriate. Optical phenomena that can be explained by these theories of light are dealt with by the so-called **physical optics**.

Huygens proposed a technique within the frame of the wave theory of light for finding the shape of a wave front at a later time from the wave front at a present time. This technique, known as **Huygens' principle** may be stated as: *each point on a wave front may be regarded as a source of secondary tiny wavelets that propagate in all directions at a speed equal to the speed of propagation of the wave. The position of the new wave front is the envelope of these secondary wavelets.*

We apply Huygens' principle to explain the propagation of light from a point light source and the phenomena of reflection and refraction. Later, we will use this principle for the explanation of the diffraction of light.

Propagation of light from a point source: It is well established that light from a point source propagates in a homogeneous isotropic medium along spherical wave fronts. Consider a point light source at S and the spherical wave front AB of light traveling at a speed υ and find the new wave front after time t (Fig. 2.12). From each point of AB we construct circles of radius $r = \upsilon t$ centered at that point. The envelope of these circles (wavelets) in the direction of propagation of the wave is the new wave front $A'B'$. It is a spherical surface as the wavefront AB.

Phenomenon of reflection: For the problem of reflection let AB be the plane wave front that impinges at an angle θ_i (angle of incidence) on a reflecting surface MN (Fig. 2.13). Let t be the time that the light travels from point B to point C with speed υ. Then $BC = \upsilon t$. The reflected wave front CD is the tangent from point C to the circle (wavelet) centered at A with radius $AD = \upsilon t$ and subtends an angle θ_r (angle of reflection) with MN. From the congruent triangles DAC and BCA we conclude that $\theta_r = \theta_i$, which is the law of reflection.

Phenomenon of refraction: For the problem of refraction let AB be the plane wave front that impinges at an angle θ_1 (angle of incidence) on the reflecting surface MN (Fig. 2.14). Let t be the time that the light travels from point B to point D on the surface MN with speed υ_1. Then $BD = \upsilon_1 t$. At the same time t the refracted wave front travels in the second medium at distance $AC = \upsilon_2 t$. The new wave front CD after time t is the tangent of the circles (wavelets) centered at points B and A with

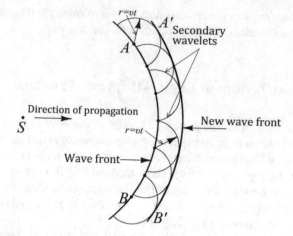

Fig. 2.12 Huygens' principle applied to wavefront AB to construct the new wavefront $A'B'$ after time t. The wave propagates with speed υ

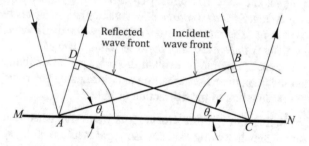

Fig. 2.13 The law of reflection explained using Huygens' principle. Incident and reflected wavefronts are perpendicular to the corresponding rays

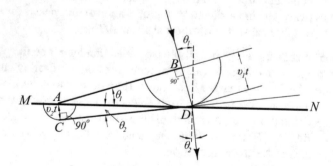

Fig. 2.14 The law of reflection explained using Huygens' principle

radii $BD = v_1t$ and $AC = v_2t$, respectively. From triangles ACD and ABD we obtain:

$$\sin \theta_1 = \frac{v_1t}{AD} \quad \sin \theta_2 = \frac{v_2t}{AD}$$

Then, since $v_1 = c/n_1$, and $v_2 = c/n_2$, we obtain $n_1 \sin \theta_1 = n_2 \sin \theta_2$, which is the law of refraction.

2.8 Electromagnetic Theory of Light

According to electromagnetic (EM) theory, light is electromagnetic radiation that is visible by the human eye and causes the sensation of vision. The EM theory can explain the phenomena of polarization, interference, and diffraction of light. All forms of EM radiation propagate in vacuum with the same **speed** of $c = 3 \times 10^8$ m/s. EM radiation is classified by **wavelength**, λ, and **frequency**, f. Wavelength, measured in meters (m), is the distance between consecutive corresponding points of the same phase on the wave, such as two adjacent crests or troughs. Frequency, f, is the number of occurrences of a periodic event per unit of time. It is measured in Hz (1 Hz = 1 cycle per second = 1 s^{-1}). The wavelength and frequency are related to the speed c of the wave by

$$\lambda f = c \qquad (2.10)$$

The EM radiation has no upper or lower limits. The various forms of EM radiation are classified by the wavelength or frequency into radio, microwave, infrared, the visible region, ultraviolet, X-rays, and gamma rays (Fig. 2.15). The wavelength of visible light is a small band of the EM spectrum. It ranges from deep red, approximately 750 nm ($f = 4 \times 10^{14}$ Hz), to violet, approximately 400 nm ($f = 7.5 \times 10^{14}$ Hz) (1 nm $= 10^{-9}$ m). The eye interprets the different wave lengths as different colors (Table 2.1).

White light generated by various sources (sun, fluorescent light bulbs, white LEDs) is defined as the complete mixture of all the wavelengths of the visible spectrum of light. A light of a specific wavelength is called **monochromatic** (from the Greek words mono, meaning one, and chromatic meaning related to color). Absolutely monochromatic light is an idealization. Light from laser is nearly monochromatic.

In the EM theory, the propagation of an EM wave is governed by Maxwell's equations. Two quantities are of great importance in EM theory: the **electric vector,** E, and the **magnetic vector**, H. It can be proven that *for wave propagation in an isotropic medium the electric vector and the magnetic vector are at right angles to the direction of the wave propagation and lie in the plane of the wave front (which is perpendicular to the direction of the wave propagation).* This means that the electromagnetic wave is a transverse wave. Furthermore, it can be proven that *the*

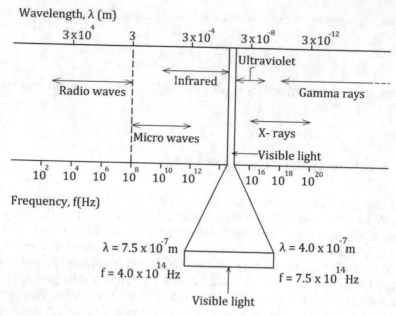

Fig. 2.15 Electromagnetic (EM) spectrum

Table 2.1 The visible spectrum

Wavelength range (nm)	Color
400–450	Violet
450–480	Blue
480–510	Blue-green
510–550	Green
550–570	Yellow-green
570–590	Yellow
590–630	Orange
630–700	Red

electric and magnetic vectors are mutually perpendicular and there is a definite ratio between the magnitudes of E and H. Both, the electric and the magnetic fields propagate with the same velocity which is the velocity of propagation of the EM wave.

From the above arguments, it is concluded that an EM wave propagating in an isotropic medium may be represented by the electric or the magnetic field vector. In the following, we will choose as the *representative vector of the EM wave the electric vector.*

The simplest form of an electromagnetic wave that propagates along the z-axis can be described by a simple sinusoidal form, as

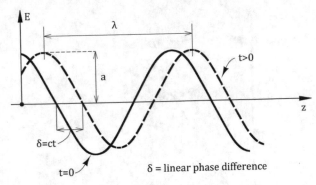

Fig. 2.16 Magnitude of the light electric vector E along the direction of propagation for two different times $t = 0$ and $t > 0$. The linear phase difference $\delta = ct$, where c is the speed of light

$$E = a \cos \frac{2\pi}{\lambda}(z - ct) \qquad (2.11)$$

where E is the magnitude of the electric vector, a is the amplitude of the wave, λ is the wavelength, c is the velocity of the wave (light) and t is the time.

Equation (2.11) expresses the magnitude of the wave E as a function of position z along the z-axis at a single instance of time $t = t_0$, as it would be seen by an observer looking along a direction normal to the direction of propagation of light. It is assumed that the light is linearly polarized, that is the electric vector lies in a single plane. For a given position $z = z_0$ along the z-axis Eq. (2.11) defines the values of the electric vector E as a function of time t. For an observer looking along the direction of light propagation toward the light source (z-direction), the wave can be represented by a line of length equal to twice the amplitude of the light vector. This corresponds to a time-exposure photograph taken by a camera aimed at the light source. Linearly polarized light is usually represented by a straight line of amplitude equal to twice the magnitude of the electric field.

Figure 2.16 represents the variation of the magnitude of the light electric vector E with the position z for two different times, $t = 0$ and $t > 0$. Note that at time t the light wave propagated through length ct. The wavelength λ is defined as the length between two corresponding points along the magnitude curve. The time required for light to move a distance of a wavelength λ is defined as the **period** T of the wave. T is given by

$$T = \frac{1}{f} = \frac{\lambda}{c} \qquad (2.12)$$

For sinusoidal wave representations we use the terms of **angular frequency**, ω, and **wave number**, ξ, given by

$$\omega = \frac{2\pi}{T} = 2\pi f, \xi = \frac{2\pi}{\lambda} \qquad (2.13)$$

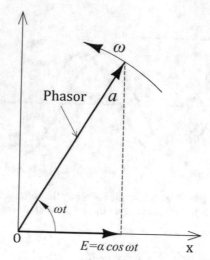

Fig. 2.17 A phasor diagram. The phasor vector with amplitude a rotates counterclockwise with a constant angular velocity ω

Then Eq. (2.11) takes the form

$$E = a \cos(\xi z - \omega t) \tag{2.14}$$

In dealing with sinusoidal equations of the form (2.11) or (2.14) the concept of the **phasor** is important. The magnitude of the electric field vector E in Eq. (2.11) or Eq. (2.14) can be considered as the projection on the x-axis of a vector of constant amplitude a which rotates counterclockwise with a constant angular velocity ω. The rotating vector is called phasor and the corresponding diagram **phasor diagram** (Fig. 2.17). The concept of the phasor is useful in adding sinusoidal quantities, as we will see later on in the study of interference and diffraction phenomena of light.

2.9 Polarization

Ordinary or unpolarized light is light in which the electric vector moves irregularly in space and does not have any directional preference. When some order is introduced into the motion of the electric vector, the light is called **polarized**. The end point of the electric vector in polarized light moves along well-defined simple curves in a definite direction. Polarized light is the simplest form of light. Unpolarized light can be defined as a mixture of all kinds of polarized light in an arbitrary and unsystematic manner.

Strictly speaking, perfectly polarized light does not exist in nature. In all cases, some amount of unpolarized light is also present. This light is called **partially polarized**. When dealing with polarized light we will consider only the ideal case of perfectly polarized light. The effect of presence of unpolarized light is only to weaken the contrast of the obtained optical patterns.

The various forms of polarized light are defined by the type of curve along which the end point of the electric vector moves. The simplest types of curves are the straight line, the circle, and the ellipse, defining the linear, circular, and elliptical forms of polarization. In the **linearly polarized light**, the end point of the electric vector moves along a straight line. The electric vector does not change direction with time, but only magnitude. We represent the light by the amplitude of the electric vector along the direction of polarization. Usually, the amplitude of the electric vector is taken equal to unity. The linear type of polarized light includes a one-parameter family, each form of linear polarization being defined by the inclination of the light vector to the horizontal.

In the **circularly polarized light**, the end point of the electric vector moves along the circumference of a circle. The magnitude of the electric vector remains constant, while its inclination varies continuously between 0 and 2π. The form of the light wave can be represented at a single instant of time by a cylindrical helix. Its transverse sectional pattern, as it is seen by an observer looking along the line of propagation of the wave toward the light source, is a circle. By defining the amplitude of the wave as being equal to unity, two different forms of circular polarization can be distinguished: the **right-circular polarization** and the **left-circular polarization**. The right-circular polarization is defined by the clockwise motion of the electric vector of light as it is seen by an observer looking along the direction of light propagation toward the light source; a counterclockwise motion corresponds to a left-circular polarization.

Elliptical polarization is the most general form of polarized light. The electric vector moves along an ellipse so that the magnitude and the orientation of the vector vary continuously. The instantaneous view of the light wave can be represented by an elliptical helix, while its transverse section by an ellipse. As in the case of circular polarization, right- and left-elliptical forms of polarization are distinguished.

Linearly polarized monochromatic light is represented by Eq. (2.14). This equation using complex variable notation can be written as

$$E = a \, Re \, e^{(\xi z - \omega t)i} \qquad (2.15)$$

where Re represents the real part of a function and i is the imaginary unit number $(i = \sqrt{-1})$.

In the most general case of elliptical polarization the electric vector E can be put in the form

$$E = E_x i + E_y j \qquad (2.16)$$

with

$$E_x = a_x \, Re \, e^{(\xi z - \omega t + \delta_x)i} = a_x \cos(\upsilon + \delta_x)$$

$$E_y = a_y \, Re \, e^{(\xi z - \omega t + \delta_y)i} = a_y \cos(\upsilon + \delta_y)$$

where $\upsilon = \xi z - \omega t$, and δ_x and δ_y are the phases of the components E_x and E_y.

A vectoral representation of polarized light can be made using the **Jones calculus**. Each form of polarized light is represented by a **Jones vector**, which is a two-element complex column vector E expressed by

$$E = \begin{bmatrix} E_x \\ E_y \end{bmatrix} = \begin{bmatrix} a_x e^{i(\upsilon + \delta_x)} \\ a_y e^{i(\upsilon + \delta_y)} \end{bmatrix} \tag{2.17}$$

In cases we are not interested in the time factor $\exp(i\upsilon)$, whose magnitude is equal to unity, the Jones vector takes the form

$$E = \begin{bmatrix} E_x \\ E_y \end{bmatrix} = \begin{bmatrix} a_x e^{i\delta_x} \\ a_y e^{i\delta_y} \end{bmatrix} \tag{2.18}$$

Usually, the Jones vector is normalized by multiplying its two elements by whatever scalar quantities are needed to reduce the value of the light intensity to unity. The normalized form of Eq. (2.18) can be written

$$E = \begin{bmatrix} \cos B e^{-i\delta/2} \\ \sin B e^{i\delta/2} \end{bmatrix} \tag{2.19}$$

where $B = |\arctan a_y/a_x|$, $\delta = \delta_y - \delta_x$.

From Eqs. (2.18) or (2.19) we can obtain the Jones vector for a given polarization. For a linear horizontally polarized light we obtain the standard and the simplified vectors as

$$E = \begin{bmatrix} a_x e^{i\delta_x} \\ 0 \end{bmatrix}, \, E = \begin{bmatrix} 1 \\ 0 \end{bmatrix} \tag{2.20}$$

For a right-circularly or a left-circularly polarized light, we have the forms

$$E = \frac{1}{\sqrt{2}} \begin{bmatrix} -i \\ 1 \end{bmatrix}, \quad E = \frac{1}{\sqrt{2}} \begin{bmatrix} i \\ 1 \end{bmatrix}, \tag{2.21}$$

The intensity of the light described by Eq. (2.18) is given by

$$I = a_x^2 + a_y^2 \tag{2.22}$$

Equation (2.22) can be put in the form

$$I = \begin{bmatrix} a_x e^{-i\delta_x} & a_y e^{-i\delta_y} \end{bmatrix} \begin{bmatrix} a_x e^{i\delta_x} \\ a_y e^{i\delta_y} \end{bmatrix} = \tilde{\mathbf{E}} \ \mathbf{E} \tag{2.23}$$

where $\tilde{\mathbf{E}}$ is the so-called Hermitian conjugate matrix of matrix $\mathbf{E}$.

2.10 Interference

2.10.1 Introduction

Interference and diffraction are the two pillars of most optical methods in experimental mechanics. **Interference** refers to the overlapping of two or more optical waves in space. It is governed by the **principle of superposition** which states that the response at any point and at any instant when two or more stimuli overlap is the sum of the responses caused by each stimulus individually. Functions that satisfy the principle of superposition are called **linear functions**. Thus, the electric field vector E that describes the resulting field obtained by superposition of two electric field vectors E_1 and E_2 is obtained by adding the two vectors as

$$E = E_1 + E_2 \tag{2.24}$$

Interference effects are most easily observed when the waves are sinusoidal with the same frequency. Also, the waves must be coherent, a condition that cannot be satisfied for the waves emitted by two independent sources. This is why in interferometric systems light is split from a single source to form more sources that are coherent, and therefore, can interfere. In conclusion, *in order to achieve interference, the sources must be monochromatic and coherent.*

2.10.2 Interference of Two Linearly Polarized Beams

Consider two linearly polarized light beams of the same frequency (or wavelength) whose electric vectors E_1 and E_2 are collinear. The magnitudes E_1 and E_2 of the vectors E_1 and E_2 can be expressed as

$$E_1 = a_1 \cos \frac{2\pi}{\lambda}(z_0 + \delta_1 - ct) = a_l \cos(\varphi_1 - \omega t)$$

$$E_2 = a_2 \cos \frac{2\pi}{\lambda}(z_0 + \delta_2 - ct) = a_2 \cos(\varphi_2 - \omega t) \tag{2.25}$$

where

δ_1 and δ_2 = phase lengths of wave E_1 and wave E_2 at position z_0, respectively.

φ_1 and φ_2 = phase angles of wave E_1 and wave E_2 at position z_0, respectively.
a_1 and a_2 = amplitudes of wave E_1 and wave E_2 at position z_0, respectively.

The magnitude of the resulting wave, E, according to Eq. (2.24) is

$$E = a_1 \cos(\varphi_1 - \omega t) + a_2 \cos(\varphi_2 - \omega t) = a \cos(\varphi - \omega t) \qquad (2.26)$$

where

$$a^2 = a_1^2 + a_2^2 + 2a_1 a_2 \cos(\varphi_2 - \varphi_1) \qquad (2.27a)$$

$$\tan \varphi = \frac{a_1 \sin \varphi_1 + a_2 \sin \varphi_2}{a_1 \cos \varphi_1 + \alpha_2 \cos \varphi_2} \qquad (2.27b)$$

Equation (2.26) shows that the resulting wave has the same frequency as the interfering two waves. The amplitude and the phase angle are different.

For the special case of equal amplitudes ($a_1 = a_2$) of the two interfering waves we obtain from Eq. (2.27a)

$$a = 2a_1 \cos \frac{(\varphi_1 - \varphi_2)}{2} = 2a_1 \cos \frac{\varphi}{2}$$

$$= 2a_1 \cos \frac{\pi}{\lambda}(\delta_1 - \delta_2) = 2a_1 \cos \frac{\pi \delta}{\lambda} \qquad (2.28)$$

where

$$\delta = \delta_1 - \delta_2, \varphi = \varphi_1 - \varphi_2 = \frac{2\pi}{\lambda}(\delta_1 - \delta_2) = \frac{2\pi \delta}{\lambda}$$

The intensity I of the resulting wave is proportional to the square of its amplitude a. It is given by

$$I \sim a^2 = 4a_1^2 \cos^2 \frac{\pi \delta}{\lambda} \qquad (2.29)$$

Equation (2.29) shows that the intensity, I, of the resulting wave from the interference of two waves of the same amplitude depends on the phase difference δ between the two waves. The intensity becomes maximum when: $\delta = n\lambda$ ($n = 0, \pm 1, \pm 2, \dots$). The maximum value of intensity is given by

$$I = 4a_1^2, \qquad (2.30)$$

Note that the intensity becomes maximum when the phase difference between the two waves is an integral number of wavelengths.

The intensity becomes minimum (zero) when: $\delta = [(2n + 1)/2]\lambda$ ($n = 0, \pm 1, \pm 2, \dots$)

$$I = 0, \tag{2.31}$$

Note that the intensity becomes zero when the phase difference between the two waves is an odd number of half-wavelengths.

From the above discussion, it is concluded that the resulting pattern from the interference of two collinear light waves of the same amplitude is a succession of bright and dark bands or **interference fringes**. When the light intensity is maximum ($\delta = n\lambda$ ($n = 0, \pm1, \pm2,...$)) we have **constructive interference**, and, when the light intensity is minimum (zero) ($\delta = [(2n + 1)/2]\lambda$ ($n = 0, \pm1, \pm2,...$)), we have **destructive interference**.

2.10.3 Young's Double-Slit Experiment

The famous double-slit experiment was performed by Young in 1801. It is one of the earliest experiments to reveal the wave nature of light and to measure the wavelengths for visible light. To obtain a proper light source for use in interference experiments Young directed light through a tiny slit 1 μm or so wide, so the emerging light presents the characteristics of an idealized light source (emissions from different parts of the source to be synchronized). Today, the experiment is performed using a coherent light source, such as a laser beam. Light illuminates a plate with two parallel slits S_1 and S_2, at distance d apart, and the resulting interference pattern is observed on a screen behind the plate (Fig. 2.18). The screen is placed at large distance R from the plate so that the emerging light beams from the slits can be considered parallel. The optical path difference between the light beams emerging from the two slits is

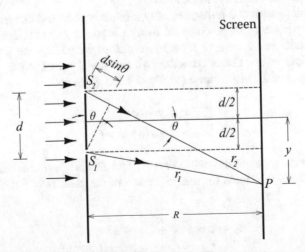

Fig. 2.18 Young's double-slit experiment. The optical path difference between the light rays emerging from the two slits S_1 and S_2 is $d \sin\theta$

$$\delta = d \sin \theta \tag{2.32}$$

where θ is the angle between a line from slits to screen and the normal to the plane of the slits.

Constructive interference on the screen occurs when

$$\delta = d \sin \theta = n\lambda \quad (n = 0, \pm 1, \pm 2, \ldots) \quad \text{[constructive interference]} \tag{2.33}$$

and destructive interference when

$$\delta = d \sin \theta = [(2n + 1)/2]\lambda \quad (n = 0, \pm 1, \pm 2, \ldots) \quad \text{[destructive interference]} \tag{2.34}$$

The fringe pattern consists of parallel, equidistant fringes parallel to the slits. The center of the pattern is a bright fringe ($n = 0$). The distance y_n of the nth bright fringe from the center of the fringes, for small angles θ ($\theta \approx \sin\theta \approx \tan\theta$), using Eq. (2.33) is

$$y_n = R \tan \theta = R\frac{m\lambda}{d} \tag{2.35}$$

and the distance D between bright (or dark) fringes is

$$D = R\frac{\lambda}{d} \tag{2.36}$$

Equation (2.36) can be used for the direct measurement of the wavelength of light, λ, by measuring the quantities R, D, and d.

The previous results of interference of two linearly polarized beams and Young's double-slit experiment can be obtained using phasor diagrams. The phasor vector of the combined waves (two or more) is obtained by adding the phasor vectors of the individual waves. Let us consider two sinusoidal waves whose electric field components E_1 and E_2 of the same amplitude a are given by

$$E_1 = a \cos \omega t$$
$$E_2 = a \cos(\omega t + \varphi) \tag{2.37}$$

The phasor diagrams of the two waves E_1 and E_2 and their combination E_p are given in Fig. 2.19. We obtain for the projection of the phasors of the two waves on the x-axis

$$E_p = a \cos \omega t + a \cos(\omega t + \varphi)$$
$$= 2a \cos(\varphi/2) \cos(\omega t + \varphi/2)$$

and by omitting the cosines term which contains the time t we obtain

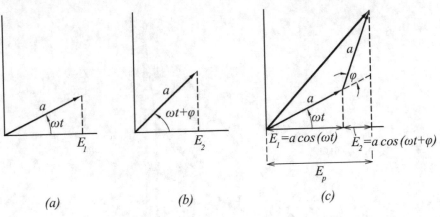

Fig. 2.19 The phasor diagrams of the two waves E_1 (**a**) and E_2 (**b**) of the same amplitude a and their combination E_p (**c**)

$$E_p = 2a \cos(\varphi/2)$$

which is Eq. (2.28).

2.10.4 Multi-slit Interference

Interference caused by more than two slits can be analyzed using phasor diagrams (Fig. 2.20). For three slits the electric field components at a point on the screen are given by

$$E_1 = a \cos \omega t$$
$$E_2 = a \cos(\omega t + \varphi)$$
$$E_3 = a \cos(\omega t + 2\varphi)$$

where φ is the phase difference between waves from adjacent slits.

The phasor diagrams for $\varphi = 0, \pi/3, 2\pi/3, \pi$ are shown in Fig. 2.21. The resultant magnitude of the electric field becomes maximum when the three phasor vectors are aligned along the same line at $\varphi = 0, \pm 2\pi, \pm 4\pi, \ldots$. These points are called *primary maxima* (Fig. 2.21a). Between the primary maxima, we find *secondary maxima* at $\varphi = \pm \pi/3, \pm \pi, \ldots$ (Fig. 2.21b, c). For these points, only one slit contributes to the resultant, while the waves from the other two slits cancel out. Destructive interference occurs whenever the three phasors form a closed triangle at $\varphi = \pm 2\pi/3, \pm 4\pi/3, \ldots$ (Fig. 2.21c). Note that the primary maxima are nine times the secondary maxima (intensity is analogous to the square of the amplitude). For n slits, the intensity of the primary maxima is n^2 the intensity of the secondary maxima. The number of

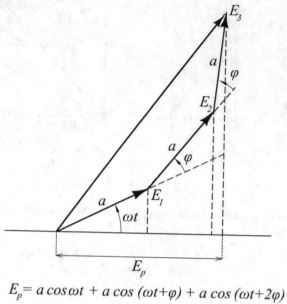

$$E_p = a\cos\omega t + a\cos(\omega t+\varphi) + a\cos(\omega t+2\varphi)$$

Fig. 2.20 A phasor diagram for the analysis of three-slits experiment

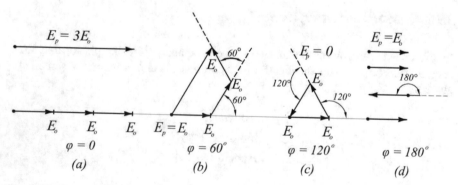

Fig. 2.21 Phasor diagrams for three-slits experiment with $\varphi = 0°$ (**a**), $\varphi = 60°$ (**b**), $\varphi = 120°$ (**c**), $\varphi = 180°$ (**d**)

secondary maxima is $n - 2$. We will study the case of very large number of slits in the diffraction gratings.

2.10.5   *Interference of Two Plane Waves*

Consider two plane waves of the same frequency and amplitude which propagate in vacuum (Fig. 2.22). The wave fronts of the two waves form an angle 2α. Let k_1 and k_2 be the wave vectors of the normal of the wave fronts. By generalizing the wave Eq. (2.15) we obtain for the electric vectors E_1 and E_2

$$E_1 = E_0 e^{i(\omega t - k_1 \cdot r)}$$
$$E_2 = E_0 e^{i(\omega t - k_2 \cdot r)}$$

where E_0 is a vector specifying the amplitude and plane of the wave, *dot* denotes the scalar product of two vectors, and vector r is given in a Cartesian coordinate system by

$$r = xi + yj + zk$$

The electric field vector E of the resulting wave is the sum of the electric field vectors of the individual waves as

$$E = E_0 e^{i\omega t}\left(e^{-ik_1 \cdot r} + e^{-ik_2 \cdot r}\right)$$

The intensity of the resulting interference pattern is proportional to the product of E and its conjugate E^*, as

$$I \sim EE^* = E_0^2\left(2 + e^{ir \cdot (k_2 - k_1)} + e^{-ir \cdot (k_2 - k_1)}\right)$$
$$= 2E_0^2[1 + \cos(r \cdot (k_2 - k_1))]$$

Let us now consider the scalar product in the argument of *cos* in the above equation by taking the z-axis to lie in the plane and bisect the angle between the unit vectors

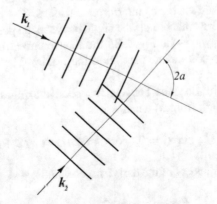

Fig. 2.22   Wavefronts of two plane waves interfering at an angle 2α

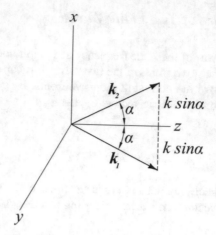

Fig. 2.23 The z-axis bisects the angle between the unit vectors k_1, k_2

k_1, k_2 (Fig. 2.23). We have

$$r \cdot (k_2 - k_1) = x(k_{2x} - k_{1x}) + y(k_{2y} - k_{1y}) + z(k_{2z} - k_{1z})$$
$$= x(k_{2x} - k_{1x}) = 2kx \sin \alpha$$

because in this coordinate system chosen, the y components of the wave vectors of the two waves are zero and their z components are identical ($k_{2y} = k_{1y} = 0$, $k_{2z} = k_{1z}$) and the difference of the x components of the two wave vectors is $2k\sin\alpha$, where the magnitude of the wave vector is $k = 2\pi/\lambda$.

The light intensity I is then

$$I \sim 2E_0^2[1 + \cos(2kx \sin \alpha)] = 4E_0^2 \cos^2(kx \sin \alpha) \tag{2.38}$$

Equation (2.38) shows that the light intensity I vary sinusoidally in space. The maxima and minima are visible as interference fringes. The plane of the fringes is perpendicular to the plane that bisects the angle between the planes of the wave fronts of the two interfering waves (in Fig. 2.23 the interference fringes form on a plane perpendicular to the z-axis and are parallel to the y-axis). The graph of the intensity distribution is shown in Fig. 2.24.

Let us calculate the distance between successive fringes. The light intensity becomes minimum (dark fringes) when

$$\cos^2(kx \sin \alpha) = 0, \quad \text{or} \quad kx \sin \alpha = \pi/2 + n\pi$$

For the positions x_1 and x_2 of the dark fringes n and $n + 1$ we obtain

$$kx_1 \sin \alpha = \pi/2 + n\pi$$

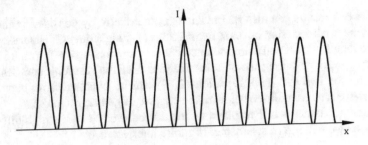

Fig. 2.24 Intensity distribution of the interference of two plane waves

$$kx_2 \sin \alpha = \pi/2 + (n+1)\pi$$

These equations render

$$k(x_1 - x_2) \sin \alpha = k\Delta x \sin \alpha = \pi$$

For $k = 2\pi/\lambda$ we obtain for the distance between two successive fringes

$$\Delta x = \frac{\lambda}{2 \sin \alpha} \tag{2.39}$$

Equation (2.39) suggests that the interference pattern gets more sparse with longer wavelengths and gets denser with increasing angles between the two waves. Table 2.2 presents values of the fringe spacing f for various values of angle α for a He–Ne laser of red light with $\lambda = 633$ nm. Note that for two perpendicular interfering waves ($\alpha = 45°$) the spatial frequency f (lines per length) of the fringes is 2234 lines/mm.

Table 2.2 Spatial fringe frequency f for various angles α of intersecting beams for a He–Ne laser of red light with $\lambda = 633$ nm

Angle α (°)	Spatial frequency f (Fringes/mm)
1	55
10	549
20	1,080
30	1,580
40	2,031
45	2,234
50	2,420
60	2,736
70	2,969
80	3,112
90	3,160

The above results can also be obtained geometrically by considering the wave fronts of the two interfering waves as two gratings of straight lines at distances equal to the wave length. Constructive interference patterns are obtained as the moiré fringes of the two gratings (along the shortest diagonals of the resulting parallelograms formed by the two gratings). Equation (2.39) can be obtained from geometrical considerations, as we will see below in the chapter of geometric moiré.

Interference patterns of two plane waves can be used to create high diffraction gratings, as we will see in the chapter of moiré interferometry.

2.10.6 Change of Phase upon Reflection—Thin Films

It can be proven theoretically (from Maxwell's equations) and verified experimentally that at normal incidence, a *light beam reflected by a material with an index of refraction greater than the index of refraction of the material in which it travels undergoes a half-cycle phase shift ($\varphi = \pi$, or $\delta = \lambda/2$) during its reflection.* Consider a light beam traveling in a material with index of refraction n_a and reflected by a material with index of refraction n_b. The amplitude of the reflected wave E_r is given by

$$E_r = \frac{n_a - n_b}{n_a + n_b} E_i \qquad (2.40)$$

where E_i is the amplitude of the incident wave.

Using Eq. (2.40) we find that for light traveling in air ($n_a = 1$) and reflected at a glass surface ($n_b = 1.5$), $E_r = -0.2E_i$. The reflected light undergoes a phase shift of $\varphi = \pi$ and has an intensity equal to 4% of the intensity of the incident light (the intensity is proportional to the square of the amplitude).

We can distinguish three cases:

i.   When $n_a > n_b$, there is no phase shift of the reflected wave relative to the incident wave. E_r and E_i have the same sign.
ii. When $n_a < n_b$, there is a half-cycle phase shift of the reflected wave relative to the incident wave. E_r and E_i have opposite signs.
iii. When $n_a = n_b$, there is no reflected wave ($E_r = 0$).

A concept frequently used in interferometry is the **optical path length** (OPL). It is defined as the product of the geometrical length times the index of refraction. For light traveling a distance d in a medium of constant refraction index n the OPL is given by

$$OPL = nd \qquad (2.41)$$

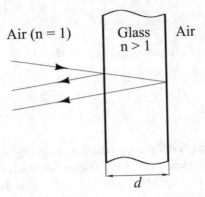

Air (n = 1)   Glass n > 1   Air

d

Fig. 2.25 A glass plate of uniform thickness d and index of refraction n in air illuminated by a light beam at normal incidence to the glass surface

When light travels in various media the optical path is the sum of the OPLs in the media. **Optical path length difference** is the difference between optical paths taken by identical waves.

As an application consider a glass plate of uniform thickness d and index of refraction n in air illuminated by a light beam of wavelength λ (in air) at normal incidence to the glass surface (Fig. 2.25). Let us consider the light rays reflected from the front and rear surfaces of the plate. The ray reflected from the front surface undergoes a phase change of $\lambda/2$ with respect to the incident wave (light travels in air ($n = 1$) and reflects at the glass surface, $n > 1$). The light reflected from the rear surface travels an extra optical path of $2dn$ and undergoes no phase change at reflection (light passes twice through the plate thickness and before reflected from the rear surface of the plate travels in glass with a higher index of refraction than air). The difference in the optical path length (DOPL) between the two rays is

$$\text{DOPL} = 2dn - \lambda/2$$

Constructive interference (bright fringes) takes place when

$$2dn - N/2 = m\lambda \text{ or } 2dn = (m + 1/2)\lambda, m = 0, \pm1, \pm2, \ldots$$

and destructive interference (dark fringes) when

$$2dn = m\lambda, m = 0, \pm1, \pm2, \ldots$$

The above equations can be put in the form:

$$2d = (m + 1/2)\lambda_n, m = 0, \pm1, \pm2, \ldots$$

and

$$2d = m\lambda_n, m = 0, \pm 1, \pm 2, \ldots$$

where $\lambda_n = \lambda/n$ is the wavelength of light in the glass plate.

2.10.7 Dispersion

When light passes through different media the frequency f of the light does not change. The boundary condition at the interface dictates that the number of wave cycles per unit time (frequency) arriving at the interface must be equal to the number of wave cycles per unit time leaving the interface. Since for a given material the product of frequency f and wave length, λ, is equal to the speed of light in the material, υ, ($\lambda f = \upsilon$) the wavelength in the material must change. If λ_0 is the wavelength in vacuum and c is the speed of light in vacuum we have that $f = \upsilon/\lambda = c/\lambda_0$, and, therefore

$$\lambda = \lambda_0 \frac{\upsilon}{c} = \frac{\lambda_0}{n} \tag{2.42}$$

where $n = c/\upsilon$ ($n > 1$) is the index of refraction of the material.

Equation (2.42) suggests that the wavelength of light in a material is smaller than the wavelength in vacuum. The wavelength becomes shorter if the wave speed decreases and becomes longer if the wave speed increases.

White light consists of all wavelengths of the visible spectrum. All wavelengths have the same speed in vacuum. However, different wavelengths have different speeds in a material. The dependence of the speed of light (and therefore the index of refraction) in the material on wavelength is called **dispersion**. Generally, in most materials, the index of refraction decreases with increasing wavelength. Because the index of refraction is greater for the shorter wavelengths, violet light is bent the most by a prism and red light the least.

2.11 Diffraction

2.11.1 Introduction

The general problem of diffraction may be stated as: *An aperture is illuminated by a light wave. Determine the amplitude of the wave at a point on a screen placed at some distance behind the aperture.* The solution of the problem in its general form is formidable. When the light source and the screen where the image is formed are close to the aperture the problem is referred to as **Fresnel diffraction**. When the light source, the aperture, and the screen are far enough away so that the rays from the source to the aperture and from the aperture to the screen can be considered parallel, the phenomenon is called **Fraunhofer diffraction**.

In the following, we will study the diffraction from a single slit, multiple slits, and a diffraction grating. We will refer to an aperture as a Fourier transform device of spatial signals and present the important function of a lens to materialize this transform within the laboratory dimensions. We will use Huygens's principle and the phenomenon of interference. At this point, we should emphasize that it is not easy to define the difference between interference and diffraction. There is no specific physical difference between them. It is fitting to quote what the famous physicist Richard Feynman wrote: *"No-one has ever been able to define the difference between interference and diffraction satisfactorily. It is just a question of usage, and there is no specific, important physical difference between them. The best we can do, roughly speaking, is to say that when there are only a few sources, say two, interfering, then the result is usually called interference, but if there is a large number of them, it seems that the word diffraction is more often used."*

2.11.2 Single Slit Diffraction

A parallel beam of monochromatic light passes through a narrow orthogonal slit of width D (Fig. 2.26a). We will derive the image of the slit on a screen placed at large distance from the slit using Huygens' principle and interference. Due to the large distance of the screen from the slit, the rays emerging form the slit can be considered parallel. Each point of the slit can be considered as a source of waves. Consider the rays that pass straight through the slit. They are all parallel and in phase, so they interfere constructively and form on the screen a bright spot.

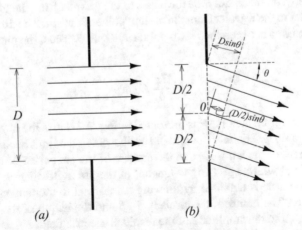

Fig. 2.26 Diffraction of light passing through a narrow rectangular slit. The long sides of the slit are perpendicular to the figure

Consider the parallel light rays that pass from the slit at an angle θ and examine the interference of two light rays from the top and the mid-point of the slit (Fig. 2.26b). The optical path length difference between these two rays is $(D/2) \sin \theta$. Each ray on the upper half part of the slit has a corresponding ray on the lower half part of the slit. All these pairs of rays have the same optical path length difference of $(D/2) \sin \theta$ as the previous two rays. They interfere on the screen and form a pattern of dark and bright bands. The condition of formation of dark bands is given by

$$\frac{D}{2} \sin \theta = m \frac{\lambda}{2}$$

or

$$D \sin \theta = m\lambda \quad (m = \pm 1, \pm 2, \ldots)[\text{destructive interference}] \tag{2.43}$$

Equation (2.43) looks similar to Eq. (2.33) for double-slit interference. However, Eq. (2.33) refers to constructive interference (bright fringes), while Eq. (2.43) refers to destructive interference (dark fringes). In Eq. (2.33) d is the distance between the two slits, while in Eq. (2.43) D is the width of the slit.

In order to find the intensity distribution on a screen placed at some distance from the plane of the slit, we can use the phasor method of adding the electric vectors of the individual wavelets emitted from the various points of the slit. Let us consider a point P on the screen at an angle θ with the normal to the slit plane (Fig. 2.27a). The total electric vector E_p at P is the vector sum of the electric field vectors of light emitted from the individual points of the slit. At the center O of the screen, all individual wavelets are in phase, and they interference constructively. Let E_0 denote the resultant amplitude of the wavelets at point O. The wavelets coming from different points of the slit travel different path lengths, and, therefore, have phase differences between them. Let β be the phase of the wavelet from the top point of the slit with respect to the wavelet from the bottom point of the slit. The difference in the optical path lengths between these two points is: $D \sin \theta$. Thus, the phase difference is:

$$\beta = \frac{2\pi}{\lambda} D \sin \theta \tag{2.44}$$

The phasor diagram at point P is presented in Fig. 2.27b. As the number of strips in the slit increases indefinitely the trail of the phasors in the limit becomes an arc of circle. The length of the arc is E_0 and the tangent of the arc at one of its end points makes an angle β with respect to the tangent of the arc at its other end point. The center C of the circle is obtained by drawing the normal to the tangents at the end points of the arc. The radius of the circle is E_0/β and the length of the chord of the arc which is equal to the amplitude of the resultant electric field, E_P, is given by

$$E_P = E_0 \frac{\sin(\beta/2)}{\beta/2} \tag{2.45}$$

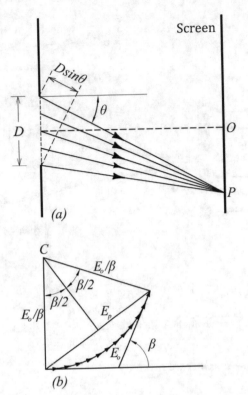

Fig. 2.27 Analysis of diffraction by a narrow rectangular slit using phasor diagram by subdividing the slit into infinitely many strips

The intensity I_0 of light at point P on the screen is proportional to the square of its amplitude E_P. It is given by

$$I = I_0 \left[\frac{\sin(\beta/2)}{\beta/2} \right]^2 \tag{2.46}$$

where $I_0 (= E_0^2)$ is the intensity at the center point O of the screen.

A plot of Eq. (2.46) is shown in Fig. 2.28 which presents the variation of the intensity I versus the angle θ. Observe that most of the power of the image is concentrated around the first maximum until the first minimum (dark fringe) at $\sin \theta = \lambda/D$. Note also that the peak intensities decrease rapidly as we move away from the center of the pattern. The dark fringes of the pattern are obtained for $\beta = 2\pi m$ ($m = \pm 1, \pm 2, \ldots$). The approximate maxima (bright fringes) of I (by putting the $\sin (\beta/2)$ term in Eq. (2.46) equal to 1) are obtained for

$$\beta = \pm(2m + 1)\pi \quad (m = 0, 1, 2, \ldots)$$

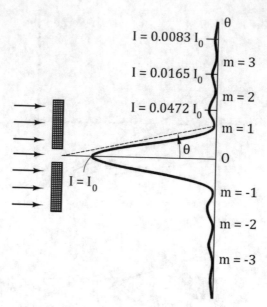

$I = 0.0083\,I_0$

$m = 3$

$I = 0.0165\,I_0$

$m = 2$

$I = 0.0472\,I_0$

$m = 1$

0

$I = I_0$

$m = -1$

$m = -2$

$m = -3$

Fig. 2.28 Intensity versus angle in single-slit diffraction

These approximate maxima are given by

$$I_m = \frac{I_0}{\left(m + \frac{1}{2}\right)^2 \pi^2} \tag{2.47}$$

where I_0 is the intensity of the central maximum and I_m is the intensity of the mth side maximum.

From Eq. (2.47) we obtain for the approximate maxima of I

$$I_1 = 0.0450\,I_0(m = 1),\ I_2 = 0.0162\,I_0(m = 2),\ I_3 = 0.0083\,I_0(m = 3)$$

The actual side maxima are:

$$I_1 = 0.0472\,I_0(m = 1),\ I_2 = 0.0165\,I_0(m = 2),\ I_3 = 0.0083\,I_0(m = 3)$$

These values indicate that the side maxima decrease rapidly and that the intensity at the first side maximum is less than 5% of the intensity of the central maximum.

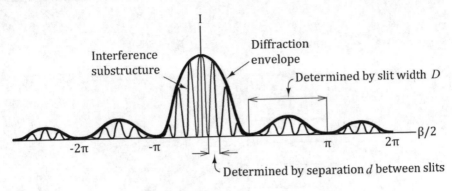

Fig. 2.29 Intensity versus angle in double-slit diffraction

2.11.3 Two-Slit Diffraction

Consider two slits of width D at a distance d. If the slits are narrow with respect to the wavelength the light intensity on the screen can be predicted from Young's two-slit interference. We obtain from Eqs. (2.29) and (2.32) for the light intensity

$$I = I_0 \cos^2\left(\frac{\pi d \sin\theta}{\lambda}\right) \qquad (2.48)$$

The interference pattern described by Eq. (2.48) consists of a series of equally spaced fringes having equal maxima. However, when the slits are not narrow and have finite width we should also include a diffraction pattern due to the individual slit. The intensity of the pattern is the product of the functions given by Eqs. (2.48) and (2.46), as

$$I = I_0 \cos^2\left(\frac{\pi d \sin\theta}{\lambda}\right)\left[\frac{\sin(\pi D \sin\theta/\lambda)}{\pi D \sin\theta/\lambda}\right]^2 \qquad (2.49)$$

The first term in the above equation is referred to as the "*interference factor*" and the second term as the "*diffraction factor*." The first term yields the interference substructure, while the second term sets limits on the number of the interference peaks (Fig. 2.29).

2.11.4 The Diffraction Grating

A diffraction grating consists of a large number N of very narrow parallel slits of the same width equally separated from each other by a constant distance d (Fig. 2.30). The spacing between the centers of adjacent slits is called the **pitch** of the grating. The number of pitches per unit length is the **spatial frequency** of the grating. The

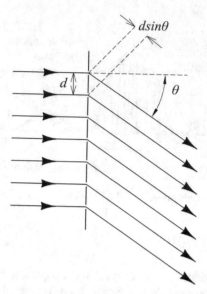

Fig. 2.30 A transmission diffraction grating. The pitch of the grating is d. Parallel light rays are incident on the grating and diffracted at angle θ

diffraction grating that contains slits is called **transmission grating**. The incident and diffracted light appear on opposite sides of the grating. The **reflection grating** is made by ruling fine lines on a surface (metallic or glass). The incident and the diffracted light appear on the same side of the grating. The light goes through in the transmission grating and it is reflected from the surface of the reflection grating. The mode of operation of both gratings is the same. Gratings that modulate the amplitude are called **amplitude gratings**, those that modulate the phase are called **phase gratings**. Amplitude gratings are low frequency gratings (10–50 lines/mm), while phase gratings are high frequency gratings (300–2400 lines/mm). The former is used in geometric moiré, the latter in moiré interferometry, and other applications of optical methods in experimental mechanics.

The analysis of a diffraction grating is based on the assumption that the slits are very narrow and spread light over a very wide angle. Also, it is assumed that the pitch of the grating d is greater than the wavelength λ. Interference on a distant screen takes place between light rays coming from the slits of the grating. Consider that the grating is illuminated by a parallel light beam. Light rays that pass though the slits without deviation interfere constructively to produce a bright fringe at the center of the pattern. Let us consider light rays at an angle θ with the normal to the slit plane. The optical path length difference between rays of adjacent slits is $d \sin \theta$. Constructive interference occurs when:

$$d \sin \theta = m\lambda \, (m = \pm 1, \pm 2, \ldots)[\text{intensity maxima}] \qquad (2.50)$$

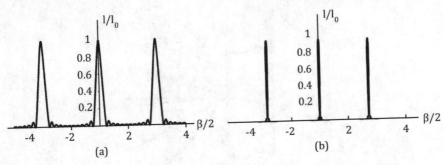

Fig. 2.31 Intensity distribution for a diffraction grating with $N = 10$ (**a**) and $N = 30$ (**b**)

The above equation is known as **grating equation**. It gives the position of intensity maxima. The number m denotes the **diffraction order** ($m = \pm 1$ for first order diffraction, $m = \pm 2$ for second order diffraction, etc.). Equation (2.50) is the same as Eq. (2.33) of the double-slit interference (Young's experiment). It indicates that maxima occur when the path difference between adjacent slits is equal to an integer number of wavelengths. The maxima for the diffraction grating occur at the same positions as for two slits with the same spacing. The location of the maxima does not depend on the number of slits. Note that for Eq. (2.50) to apply the pitch of the grating d must be greater than λ ($\sin \theta < 1$).

There is, however, a fundamental difference in the grating (N slits) and the two-slit pattern. The maxima become sharper and more intense as the number of slits increases. Furthermore, while in the two-slit pattern there is one minimum between two maxima, in the N-slit pattern there are $(N - 1)$ minima between each pair of principal maxima. A minimum occurs whenever the phase difference between adjacent slits is an integral multiple of $2\pi/N$. The height of each principal maximum is proportional to N^2, and its width is proportional to $1/N$. Figure 2.31 shows the intensity distribution as a function of $\beta/2$ for two diffraction gratings with $N = 10$ and 30. The principal maxima become sharper and more narrow for $N = 30$ than for $N = 10$.

The above behavior of gratings with large number of slits can be explained as follows: consider that the angle θ is increased slightly beyond that required for a maximum. In the case of two slits, the waves will be nearly in phase and full constructive interference takes place so that the maxima are broad. In a grating with a large number of slits, the waves from two adjacent slits will also be nearly in phase. However, waves from one slit may be exactly out of phase with waves from another slit far away. In this way, most of the interfering waves cancel out.

In conclusion, *a grating produces exactly the same pattern, dictated by Eq. (2.50), as the pattern of the two slits in Young's experiments. However, the bright fringes in the grating are sharper and brighter than in the two slits alone.* Equation (2.50) is very important and constitutes the basis of many optical methods of experimental mechanics.

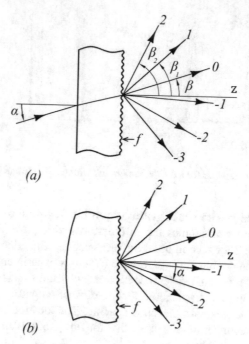

(a)

(b)

Fig. 2.32 Diffraction by a transmission (**a**) and reflection (**b**) grating at oblique incidence of light

The above analysis of a diffraction grating is the result of superposition (interference) of waves coming from many slits. It appears that a more appropriate term for the diffraction grating is "*interference grating*".

In the previous discussion, the light beam is normal to the plane of the grating. When the incident beam makes an angle α with the normal (Fig. 2.32) the angles of diffraction are determined by the following equation

$$\sin \beta_m = \sin \alpha + m\lambda f \tag{2.51}$$

where β_m is the angle of the mth diffraction order ($m = 0, \pm 1, \pm 2, \ldots$), f is the grating spatial frequency ($f = 1/d$) and λ is the wavelength of light. The angles are measured from the normal to the grating surface. Angle β_m cannot be greater the 90°. The zero diffraction order is an extension of the incident beam for transmission gratings (Fig. 2.32a) and is symmetrical with respect to the normal to the grating surface for reflection gratings (Fig. 2.32b). Equation (2.51) for $\alpha = 0$ (normal incidence of light) reduces to Eq. (2.50). Equation (2.51) is the **grating equationunder oblique incidence of light**. It is a very important equation in optical methods of experimental mechanics.

The condition that $\beta_m < 90°$ imposes restrictions on the number of diffraction orders for a grating with a specific spatial frequency. Table 2.3 presents the number of diffraction orders and the angle of first order diffraction at normal incidence for

Table 2.3 Number of diffraction orders and first order diffraction angle for various grating frequencies at normal incidence for a red Helium–Neon laser ($\lambda = 633$ nm)

f (lines/mm)	Number of refraction orders	First order diffraction angle, β_1 (°)
10	317	0.36
40	79	1.45
100	31	3.63
1,000	3	39.3
1200	3	49.4
1,500	3	71.7

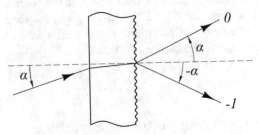

Fig. 2.33 The zeroth (0) and the -1 diffraction orders are symmetrical when $\sin\alpha = \lambda f/2$

various grating frequencies for a Helium–Neon red laser with $\lambda = 632.8$ nm. Note that for the last three grating of frequency 1,000, 1,200 and 1,500 lines/mm only three diffraction orders are obtained ($m = 0, 1, -1$).

An interesting case occurs when the zeroth and the -1 diffraction orders are symmetrical with respect to the grating normal, that is, $\beta_0 = \alpha$, and $\beta_{-1} = -\alpha$ (Fig. 2.33). Equation (2.51) gives for the zeroth and the -1 diffraction orders

$$\sin\beta_0 = \sin\alpha, \ \sin\beta_{-1} = \sin\alpha - \lambda f$$

Since $\beta_{-1} = -\beta_0$, we obtain for the angle of incidence α

$$\sin\alpha = \frac{\lambda f}{2} \tag{2.52}$$

Equation (2.52) gives the angle of incidence for which the zeroth and the -1 diffraction orders are symmetrical with respect to the grating normal.

Let us determine the conditions under which only two symmetrical diffractions emerge from the grating (Fig. 2.33). The lower limit is obtained by putting $\beta_1 = 90°$. We have from Eqs. (2.51) and (2.52).

$$f = \frac{2}{3\lambda} \text{ and } a = 19.5°$$

The upper limit is obtained for $\alpha = 90°$. Then Eq. (2.52) gives

$$f = \frac{2}{\lambda}$$

By combining the above two equations we obtain that only two symmetrical orders emerge from the grating when

$$\frac{2}{3\lambda} < f < \frac{2}{\lambda} \tag{2.53}$$

From Eq. (2.53) we obtain that for a Helium–Neon red laser $1,053 < f < 3,160$.

It is interesting to compare Eq. (2.39) which gives the distance between two successive interference fringe of two plane waves whose fronts make an angle 2α with Eq. (2.52). Equation (2.39) can be written as

$$\sin \alpha = \frac{\lambda}{2(\delta x)} = \frac{\lambda f}{2} \tag{2.54}$$

where $f = 1/(\delta x)$ is the frequency of the interference fringes.

Note that Eq. (2.54) of two-beam interference is the same as Eq. (2.52) of special diffraction in which the diffracted beam is symmetrical to the incident beam. This means that if the grating formed by the interference pattern of two beams is illuminated by one of the interfering beams the resulting non-zero diffracted order constitutes the other interfering beam. This is known as the **equivalence principle between interference and diffraction**. It is of great importance in holography.

2.11.5 Diffraction by a Circular Aperture

The diffraction pattern formed by a uniformly illuminated circular aperture consists of a bright central region surrounded by a series of concentric bright and dark rings. It is known as the **Airy disk**, in honor of Sir George Airy (1801–1892) who first derived the expression for the intensity in the pattern. Figure 2.34 presents the intensity of light across the diffraction pattern of a circular hole. The pattern is described by the angular radii of the rings. The angle θ_1 that the line from the center of the aperture to the first dark ring makes with the normal to the plane of the pattern is given by

$$\sin \theta_1 \approx \theta_1 = 1.22 \frac{\lambda}{D} \tag{2.55}$$

where D is the aperture diameter and λ is the wavelength. Note that this formula differs from that of a slit (Eq. (2.43) with $m = 1$) by the factor 1.22. This is due to the non-uniform width of the circular slit, as compared to the uniform width of the

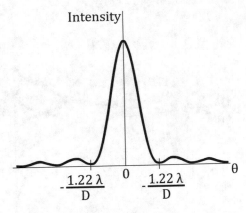

Fig. 2.34 Intensity of light across the diffraction pattern of a circular hole

rectangular slit. It can be shown that the "average" width of a circle of diameter D is $1.22D$.

The angular radii θ_2 and θ_3 of the next two dark rings are given by

$$\sin \theta_2 = 2.23 \frac{\lambda}{D}, \quad \sin \theta_3 = 3.24 \frac{\lambda}{D}$$

The angular radii of the bright rings between the dark rings are given by

$$\sin \theta = 1.63 \frac{\lambda}{D}, 2.68 \frac{\lambda}{D}, 3.70 \frac{\lambda}{D}$$

2.11.6 *Limit of Resolution*

When two point objects are very close the diffraction patterns of their images will overlap. As the two objects move closer it is not distinguishable if there are two overlapping images or a single image. The images of two point objects are resolved (they are distinguishable) when they are seen as separate. The minimum separation of two objects that can be resolved by an optical system is the **limit of resolution** of the optical system. Diffraction effects limit the resolution of a system. A generally accepted criterion for resolution of two objects is the **Rayleigh criterion**, according to which, *two images are resolved when the center of the Airy disk of one image coincides with the first dark ring of the other* (Fig. 2.35). The limit of resolution of an optical system is given by Eq. (2.55). The smaller the limit of resolution, the greater the resolving power of an optical system. The limit of resolution of the human eye for most people is 5×10^{-4} rad. The limit of resolution of the Hubble telescope for visible light with $\lambda = 550$ nm is of the order of 3×10^{-7} rad.

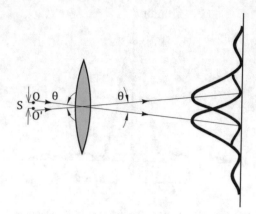

Fig. 2.35 The Rayleigh criterion. The images of point objects O and O' subtending an angle θ at the lens are resolvable when the center of the Airy disk of one image is directly over the first minimum in the diffraction pattern of the other. Only one ray, passing through the center of the lens, is drawn for each object to indicate the center of the diffraction pattern of its image

2.11.7 *Fraunhofer Diffraction as a Fourier Transform*

It can be shown that in the Fraunhofer diffraction the far field on the image plane is the Fourier transform of spatial signals (grating, lens) in an aperture. Hence, light passing through an aperture will produce the Fourier transform of the aperture plane. Given a function of transparency, its Fourier transform gives the image of transparency. In some cases, for the Fraunhofer approximation to be valid the distance from the aperture to the image plane may be very large, beyond the limits of the laboratory. In order to bring the image to laboratory dimensions, we use a lens. For this reason, *the lens is considered a Fourier transform device.* We should emphasize that it is not the lens that is the Fourier transforming devise. It is the aperture with or without a lens. The lens makes it possible to make the transform within the dimensions of the laboratory. The transform will be visible at the focal plane of the lens. The lens as a Fourier transforming device is extensively used in optical methods of experimental mechanics.

As an application consider the case of a grating whose amplitude transmittance varies sinusoidally for normal incidence of light. Since the grating has one harmonic, its Fourier transform will produce, besides the zeroth order, the $+1$ and -1 orders that deviate from the zeroth order. Thus, *a sinusoidal amplitude transmission grating produces two diffracted beams.* If the grating is not a sinusoidal one, its amplitude can be considered as a series of sinusoidal terms, and its Fourier transform will produce two inclined diffraction orders for each sinusoidal component of the grating. When the grating transmittance consists of two sine waves, the diffracted pattern will contain four inclined diffraction orders of ± 1, and ± 2.

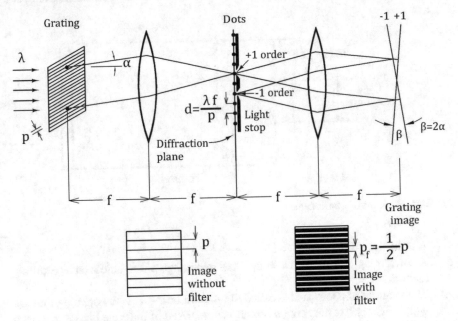

Fig. 2.36 Fourier transform of a diffraction grating

2.11.8 *Optical Spatial Filtering*

Consider in Fig. 2.36 the Fourier transform of a diffraction grating (it can be any optical signal) by a lens on an image plane. If we place a second lens it will perform the inverse transform, so that on a plane after the second lens we will obtain the same diffraction grating. We can block or modify parts of the Fourier transform so that we modify the image of the diffraction grating. This process is called **optical spatial filtering** or **optical Fourier processing**. Let us block all dots of the diffraction pattern, but the +1 and −1 orders. Then we will obtain a diffraction grating with double spatial frequency. If a filter is inserted in the diffraction plane which allows only rays from the +2 and −2 diffraction orders to pass, the frequency of the image of the diffraction grating is four times the frequency of the diffraction grating.

2.12 Camera

Camera is an optical device that makes an image of an object and records it on film or electronically. It is widely used in experimental mechanics. The word camera comes from Latin and means chamber. The basic elements of a camera are (Fig. 2.37):

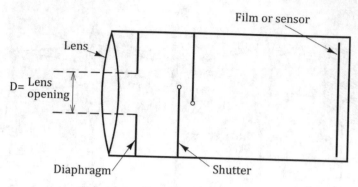

Fig. 2.37 Elements of a basic camera

a. A light-tight box.

b. A converging lens. It makes an inverted real image of the object on a recording medium.

c. The aperture of a lens is also called diaphragm or iris. It is the opening through which light enters the camera box. It is comprised of multiple blades that open and close. The aperture can be widened or narrowed to control the amount of light that enters the camera box.

d. A shutter. When the shutter is opened it allows light from the object to be focused by the lens as an image on the film or sensor.

e. A piece of film in traditional cameras or a semiconductor sensor in digital cameras.

The charge-coupled device (CCD) sensor in digital cameras is made up of millions of tiny pixels (picture elements). A 3-megapixel (3×10^6) sensor contains about 1500 pixels vertically by 2000 pixels horizontally ($1500 \times 2000 = 3 \times 10^6$) over an area of about 9 mm $\times$ 12 mm. When light reaches pixel electrons are liberated from the semiconductor. The charge of the pixel is carried out by conducting electrodes to a central processor that stores the brightness of pixels and allows reformation of the image on a computer screen. When the charges of the pixels are transferred to memory a new picture can be taken. Besides the CCD sensor, digital cameras use the CMOS (Complementary Metal–Oxide–Semiconductor) sensor. This sensor has two important characteristics, high noise immunity, and static power consumption. These characteristics allow CMOS to integrate a high density of logic functions into a chip. Both types of CCD and CMOS sensors have been used successfully in many applications of digital image correlation (DIC) method (Chap. 12).

There are three main adjustments in a camera: shutter speed, f-stop, and focusing.

Shutter speed: It refers to the time the shutter is open and the film or sensor is exposed. It may vary from a second (s) or more to 1/1000 s or less. Normally speeds faster than 1/100 s are used to avoid blurring from camera movement. For moving objects faster shutter speeds are needed.

f-stop: The amount of light that reaches the film or the electronic sensor must be controlled to avoid underexposure (little light) or overexposure (too much light). In the first case, the picture is dark and only the brightest objects appear, while in the second case all bright objects look the same with a consequent lack of contrast.

The amount of light that enters the camera depends on the focal length of the camera lens. A lens of long focal length, called a *telephoto* lens, gives a small angle of view and a large image of a distant object, while a lens of short focal length gives a wide angle of view and a small image. For any object distance using a lens of longer focal length gives a greater image distance. This behavior can be deduced from Eqs. (2.7) and (2.8). Indeed we have

$$\frac{d_i}{d_0} = \frac{1}{1 - f/d_0} = -\frac{h_i}{h_o} \tag{2.56}$$

here h_0 and d_0 are the height and the distance of the object from the lens and h_i and d_i are corresponding quantities of the image, and f is the focal length of the lens.

When f is small, the ratios d_i/d_0 and h_i/h_0 are small, and a distant object gives a small image. When f is large the image of the same object may entirely cover the area of the film.

The intensity of light that enters the camera is proportional to the area the camera lens views and the effective area of the lens. The area the lens views is proportional to the square of the angle of view of the lens, thus, approximately is proportional to $1/f^2$. Furthermore, the effective area of the lens is proportional to the square by the diameter D of the diaphragm, D^2. Thus, the intensity of light reaching the film is proportional to D^2/f^2. A number that is related to the intensity of the light is the ratio f/D, called the **f-stop** or **f-number** of the camera lens:

$$f\text{-stop }(f\text{-number}) = \frac{\text{Focal length}}{\text{Aperture diameter}} = \frac{f}{D} \tag{2.57}$$

The intensity of light reaching the film is inversely proportional to the f-number. The smaller the f-stop number more light passes through the lens. Standard f-stops markings on good lenses are: 1.0, 1.4, 2.0, 2.8, 4.0, 5.6, 8, 11, 16, 22, and 32. Each of these f-stops corresponds to a diameter reduction of the effective area of the lens by a factor of about $\sqrt{2} = 1.41$. Because the amount of light that reaches the film if proportional to D^2 each f-stop number corresponds to a factor of *two* in light intensity reaching the film.

Focusing: When an object is photographed the lens should be placed at the correct position from the film to obtain the sharpest image. The distance d_i of the image plane from the lens is dictated by the lens Eq. (2.7) as

$$\frac{1}{d_o} + \frac{1}{d_i} = \frac{1}{f} \tag{2.7}$$

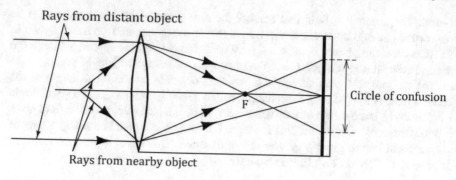

Fig. 2.38 When rays from a nearby object are focused on the film, rays from distant objects produce circles of confusion and the photo is blurred

From this equation, we obtain that for an object at infinity ($d_o \to \infty$) $d_i = f$, that is, the object is focused at the focal point of the lens. For closer objects with $d_o > f$ d_i increases as the object distance increases. Thus, the lens must be moved away from the film in order for the image of the object to be formed on the film. The operation of placing the lens at the correct position so that the image of the object is formed on the film is called *focusing*.

Consider a 50 mm-focal length camera lens and consider an object at distances from the lens: $d_o = 2$ m, 3 m, 5 m and ∞. From Eq. (2.7) we obtain $d_i = 51.3$ mm, 50.8 mm, 50.5 mm, and 50 mm, respectively. Thus in order for the object to be focused on the film the lens should be moved away from the film by 1.3 mm, 0.8 mm, 0.5 mm, and 0, respectively.

For a nearby object, its image is focused on the film by moving the camera lens away from the film. However, distant objects will be out-of-focus. The rays from these objects at far distances will form an image of overlapping circles on the film (Fig. 2.38). These circles are called **circles of confusion**. For near and distant objects to be included in a photo the lens should focus at an intermediate distance. At such position, there is a range of distances for which the circles of confusion are small that the images will be reasonably sharp. This is called the **depth of field**. For small lens openings only rays through the central part of the lens are taken and form smaller circles of confusion for a given object distance.

The sharpness of the picture depends on the graininess of the film or the number of pixels in a digital camera. The quality of the lens as dictated by the lens resolution also affects the quality of the picture. The sharpness of a lens is often given by the number of lines per mm measured by photographing a set of parallel lines. The smaller the spacing between distinguishable lines the better the resolution.

Further Readings

1. Rossi B (1965) Optics, 3rd edn. Addison-Wesley
2. Born M, Wolf E (1975) Principles of optics, 5th edn. Pergamon Press
3. Theocaris PS, Gdoutos EE (1979) Matrix theory of photoelasticity. Springer, pp 6–104
4. Dally JW, Riley WF (1991) Experimental stress analysis. McGraw-Hill, pp 343–377
5. Cloud GL (1998) Optical methods in engineering analysis. Cambridge University Press, pp 13–53
6. Giancoli DC (2005) Physics, 6th edn. Pearson–Prentice Hall
7. Young HD, Freedman RA (2007) University physics, 12th edn. Pearson-Addison Wesley
8. Sciammarella CA, Sciammarella FM (2012) Experimental mechanics of solids. Wiley, pp 123–284
9. Shukla A, Dally JW (2014) Experimental solid mechanics, 2nd edn. College House Enterprises, pp 259–282

Chapter 3
Geometric Moiré

3.1 Introduction

Geometric moiré is an optical method of experimental mechanics for measurement of displacements and slopes. It is based on the phenomenon of obstruction of light when it passes through two superposed gratings. The moiré effect can be explained by geometric optics and there is no need to invoke the wave theory of light. Moiré is the French name of textile with wavy (watered) appearance produced mainly from silk, but also wool, cotton, and rayon.

In this chapter, we develop the basic principles of the moiré phenomenon and its applications in experimental mechanics. Following the definition of basic terms, we present the moiré phenomenon, the mathematical formation of moiré fringes, the relationship between line gratings and moiré fringes, the moiré patterns formed by circular, radial and line gratings, the measurement of in-plane and out-of-plane displacements and slopes, the sharpening of moiré fringes and the moiré of moiré effect.

3.2 Terminology

The moiré phenomenon occurs whenever a repetitive structure, such as a mess, is overlaid with another such structure. The two structures need not be identical. The effect is observed in everyday life. Examples include the pattern seen through two rows of a mesh, or picket fence from a distance, layers of window screens, or coarse textiles. Before studying the moiré effect we will define a few terms that are needed for its understanding.

The basic element in the moiré method is the **grating**. It consists of equidistant opaque bars of constant width separated by transparent slits. The bars are usually straight lines, even though other forms of lines are used, like circular, radial, etc. The width of the opaque bars is usually equal to the width of the transparent slits.

© The Author(s), under exclusive license to Springer Nature Switzerland AG 2022
E. E. Gdoutos, *Experimental Mechanics*, Solid Mechanics and Its Applications 269,
https://doi.org/10.1007/978-3-030-89466-5_3

In some cases, the widths are different producing desirable results. The ensemble is made of an opaque bar and its adjacent transparent slit is termed **ruling** or **line** of the grating. The distance between corresponding points of two successive bars or slits is termed **pitch** and is usually denoted by p. The number of rulings per unit length is called **spatial frequency** of **density** of the grating. Gratings are characterized by their density. In geometric moiré, the density of the grating cannot exceed 40 lines per mm. For higher density gratings diffraction effects enter and a different approach for the interpretation of the obtained optical pattern is needed. The above gratings that consist of periodic dark lines are called **amplitude gratings**. There are gratings that transmit light over their whole surface and change the phase of the transmitted light by a periodic variation in thickness or refraction index. They are called **phase gratings**. Some gratings have the combined characteristics of phase and amplitude gratings. Both types of gratings when combined with another grating produce the moiré effect. In this chapter, we will use only amplitude gratings.

The direction perpendicular to the rulings of a grating is termed **principal direction**, while the direction parallel to the rulings is called **secondary direction**. Superposed gratings form successive bright and dark bands, called moiré fringes. The distance between two successive fringes is termed **interfringe spacing** and is denoted by f.

In moiré analysis of strain, two gratings are usually used. One, attached to the specimen, referred to as **model** or **specimen grating,** and another superposed to the specimen grating referred to as **master** or **reference grating**. Moiré fringes are produced from the superposition of the two gratings. The rulings of the gratings can be regarded as indexed families of curves. Let $S(x, y) = k$ denotes the family of curves corresponding to the specimen grating and $R(x, y) = l$ the family of curves corresponding to the reference grating. k and l are indexing parameters running over a subset of real integers $(0, \pm 1, \pm 2, \ldots)$. Their values define the spacing of the rulings of the gratings.

3.3 The Moiré Phenomenon

Consider two line gratings of the same pitch with equal width of opaque bars and transparent slits printed on transparent sheets. Superpose the gratings so that their principal directions coincide (Fig. 3.1). When light passes through the gratings a uniform light or uniform dark field is observed, depending on whether the opaque bars of one grating coincide exactly with the opaque bars of the other grating or weather the opaque bars of one grating coincide exactly with the transparent slits of the other grating, respectively. If one grating is displaced with respect to the other so that the principal directions of the two gratings coincide a uniform field of alternate light and dark bands appear whenever the movement of one grating with respect to the other is half a pitch. Between the light and dark bands, an intermediate gray field of continuously varying intensity appears. The intensity of light behind two superimposed gratings of the same pitch whose dark bars coincide is shown

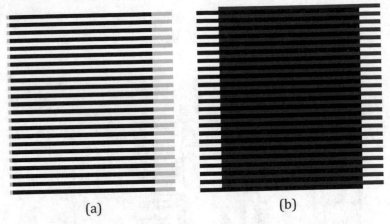

(a)   (b)

Fig. 3.1 Superposition of two line gratings of the same pitch with equal width of opaque bars and transparent slits. The opaque bars and transparent slits of both gratings coincide (bright field) (**a**), The opaque bars of one grating coincide with the transparent slits of the other (dark field) (**b**)

in Fig. 3.2. The eye blends the optical effect and sees uniform average light. The transmitted light intensity behind the gratings is 50% of the incident intensity. It is assumed that the frequency of the gratings is low (lower than 40 lines per mm) to exclude diffraction effects. The light and dark bands that occupy the whole field as one grating moves relative to the other grating are called **moiré fringes**. Note that when the two line gratings are moved along the secondary direction (parallel to their lines) no optical effect is produced.

The above observations can be summarized as: *Whenever two line grating of the same pitch are displaced one with respect to the other along their principal directions and remain parallel alternate bright and dark fringes appear when the relative movement of one grating with respect to the other equals one pitch. No optical effect is produced when the gratings move in their secondary direction and remain parallel* (the direction normal to the grating lines is called principal direction because when the gratings are moved in this direction an optical (moiré) effect is produced, while the direction parallel to the grating lines is called secondary direction because when the gratings are moved in this direction no optical effect is produced). This is the very essence of the moiré phenomenon.

The relative motion υ of the two gratings is given by

$$\upsilon = np \tag{3.1}$$

where n is the number of bright or dark cycles and p is the pitch of the gratings. Equation (3.1) constitutes the fundamental equation of the moiré phenomenon. Small displacements of one pitch are magnified by their transformation into moiré fringes.

From the above analysis, it is clear that the geometric moiré effect is produced by obstruction of light. It does not involve interference or diffraction effects. It can

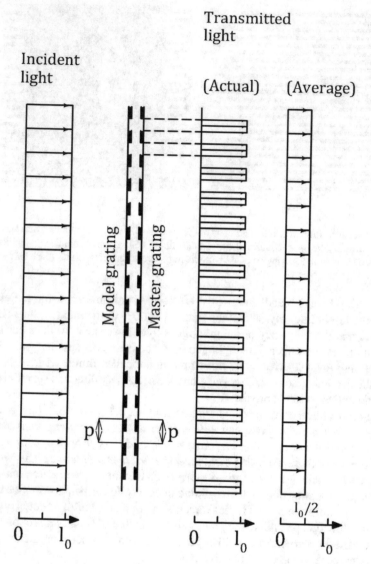

Fig. 3.2 Light intensity behind two superimposed gratings of the same pitch whose opaque slits and dark bars coincide

be explained merely by geometric optics (based on the ray model) and there is no need to invoke the wave theory of light. *It is the only optical method of experimental mechanics treated in this book, besides the method of caustics, that can be explained by geometric optics.* This is the reason that the moiré method is often referred to as **mechanical interference** in contrast to optical interference which involves the wave theory of light.

Consider the superposition of two line gratings of different pitches p and $p_1 = p(1 + \lambda)$ with lines parallel to each other (Fig. 3.3). Assume that the two gratings had originally equal pitch p and one is attached to a tension specimen. The pitch of the specimen grating increases to p_1 ($p_1 > p$, $\lambda > 0$) as the applied tensile load increases When an opaque strip of one grating falls over a transparent strip of the other grating a dark fringe appears. On the other hand, when two opaque strips coincide there is a bright fringe. Note that the direction of the fringes coincides with the direction of the grating lines. Let m be the number of lines of the master grating of pitch p that fall between two fringes at distance δ apart (Fig. 3.3). We have

$$\delta = mp = (m - 1)p_1$$

and the specimen pitch p_1 is given by

$$p_1 = p\frac{m}{m - 1} = \frac{\delta}{\frac{\delta}{p} - 1} \tag{3.2}$$

If the master grating is attached to a compression specimen the minus ($-$) sign in the above equations is replaced by the plus ($+$) sign.

Let us now define the **Eulerian strain** (the change of length divided by the final length) by

$$\varepsilon^E = \frac{l_f - l_i}{l_f} = \frac{p_1 - p}{p_1} \tag{3.3}$$

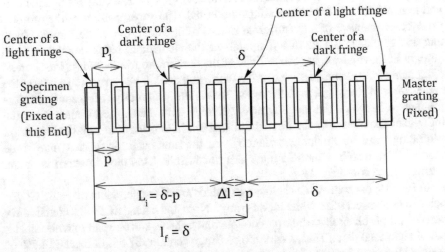

Fig. 3.3 Formation of moiré fringes by superimposing two parallel line gratings of different pitch p_1 and p

where l_i is the initial length before deformation and l_f is the final length after deformation.

From Eqs. (3.2) and (3.3) we obtain

$$\varepsilon^E = \frac{p}{\delta} \tag{3.4}$$

Equation (3.4) expresses the Eulerian strain ε^E in terms of the pitch p of the reference grating and the distance δ between two successive fringes. The relative displacement of points on one fringe with respect to points on a successive fringe is equal to the pitch of the reference grating. The total relative displacement of two points belonging to two fringes is equal to the number of fringes between the points times the pitch of the reference grating. Note that the moiré fringes are formed on the deformed positions of the body and Eq. (3.4) determines the Eulerian strain.

From Eqs. (3.3) and (3.4) we obtain for the distance δ between two successive moiré fringes

$$\delta = \frac{pp_1}{p_1 - p} = p\frac{1 + \lambda}{\lambda} \approx \frac{p}{\lambda} \quad (\lambda \ll 1) \tag{3.5}$$

Equation (3.5) indicates that δ is inversely proportional to the relative pitch difference of the two gratings $\lambda(= (p_1 - p)/p)$. For example, for $\lambda = 0.01, \delta \approx 100$ p.

The above analysis is concerned with the superposition of two line gratings of different pitches and parallel rulings. Consider now the superposition of two line gratings of equal pitch at an angle to each other (Fig. 3.4). The width of opaque bars and transparent slits for both gratings are equal. The superposition of the gratings produces rhombuses of three different shades of darkness, i.e., white, simple black, and double black. The shortest diagonals of the elementary black rhombuses form straight lines which are the center lines of the dark moiré fringes, while the shortest diagonals of the elementary successive white and double black rhombuses form the center lines of the bright moiré fringes. This is the **subtractive moirépattern**. It is the pattern in which the moiré fringes coincide with the shortest diagonals of the individual rhombuses or the interfringe spacing is longest. The same construction can be repeated for the longest diagonals of the rhombuses or shortest interfringe spacing. The resulting moiré fringes are not visible. This moiré pattern is called **additive moirépattern**.

In Fig. 3.4 the zone $ABCD$ belongs to a dark moiré fringe, while the zone $CDEF$ belongs to its successive white moiré fringe. Note that each dark fringe contains six parts of simple black shade, one part double black shade, and one part of white shade, out of eight total parts. Thus, a dark moiré fringe contains 87.5% of black (12.5% of double black and 75% of simple black) and 12.5% of white shade. A bright fringe contains two parts of simple black shade, three parts double black shade, and three parts of white shade, out of eight total parts. Thus, a bright fringe contains 62.5% of black (37.5% of double black and 25% of simple black) and 37.5% of white shade.

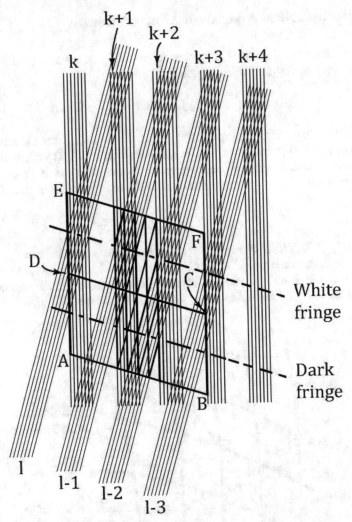

Fig. 3.4 Dark and white moiré fringes formed from the superposition of two line gratings of equal pitch with equal width of opaque bars and transparent slits at an angle to each other. Rhombuses of white, simple black, and double black shades are shown

From the above, it is concluded that in dark or bright moiré fringes all regions are neither completely black nor completely white. In a bright fringe, the white shade is three times the white shade of a dark fringe.

3.4 Mathematical Analysis of Moiré Fringes

Consider two indexed families of curves representing the specimen and reference gratings expressed by the indicial equations

$$S(x, y) = k, R(x, y) = l(k, l = \pm 1, \pm 2, \ldots)$$ (3.6)

When the two gratings are superposed the resulting moiré pattern is expressed by an indexed family of curves $M(x, y) = m$ ($m = \pm 1, \pm 2, \ldots$). The moiré fringes are the diagonals of the curvilinear quadrangles formed by the intersection of the curves of the two gratings. They are expressed by the following indicial equation (Fig. 3.5).

$$k \pm l = m$$ (3.7)

Points E, F, G, H, … of Fig. 3.5 correspond to the moiré pattern for which $(k - l) = constant$, while points A, B, C, D, … of Fig. 3.5 correspond to the moiré pattern for which $(k + l) = constant$. The moiré fringes for which the equation $(k - l) = m$ applies is called the **subtractive moirépattern**, while the moiré fringes for

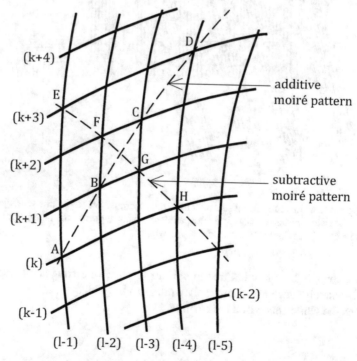

Fig. 3.5 Indicial notation of formation of additive and subtractive moiré patterns. The effective moiré pattern belongs to the subtractive type

which the equation $(k + l) = m$ applies is called the **additive moirépattern.** As we mentioned previously, the effective or visible moiré pattern is the pattern with the longest intefringe spacing or the shortest diagonals of the individual quadrangles. The effective moiré pattern in Fig. 3.5 is the pattern containing the fringe *EFGH* and belongs to the subtractive moiré pattern.

Equation (3.7) indicates that the moiré pattern is the result of adding or subtracting two functions that are expressed parametrically. The moiré pattern can also be considered as a family of parametric curves. Knowledge of any two of the three families of curves allows the determination of the third. In displacement analysis, one family is the master grating and a second is the moiré fringes. The third family is the deformed specimen grating from which the displacements of the specimen are measured.

The effective moiré pattern may change from subtractive type to additive type and vice versa. The boundary between the two moiré types is the region in which the individual quadrangles are squares. It is called **commutation moiré boundary** or simply **moiré boundary.**

Consider in Fig. 3.6 the formation of moiré fringes by two superimposed gratings. The lines $(k − 1)$, k, $(k + 1)$ of one grating and the lines l and $(l + 1)$ of the other grating are shown in the figure. The moiré fringes m and $m + 1$ are also shown. Let all curves are referred to a Cartesian system of coordinates centered at the intersection of the curves l, k, and m, and θ and φ are the angles the x-axis makes with the diagonals OO' and OO''. The interfringe spacing f_{ms} is defined by the distance OK on the normal of fringe m at point O. We obtain from triangle $OO'K$ for f_{ms}

$$f_{ms} = a \ \sin(\theta − \varphi) \tag{3.8}$$

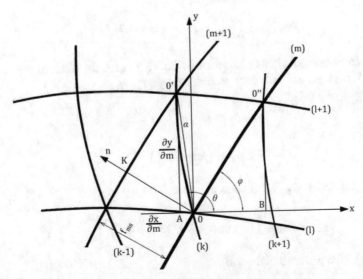

Fig. 3.6 Geometry of formation of moiré fringes by two superimposed gratings

where the distance $a = OO'$ of the diagonal between fringes m and $m + 1$ is given by

$$a = \left(x_{,m}^2 + y_{,m}^2\right)^{1/2} \tag{3.9}$$

where comma (,) denotes the derivative of the respective quantity. The angles θ and φ are expressed by

$$\tan \theta = \frac{y_{,m}}{x_{,m}}, \tan \varphi = -\frac{M_{,x}}{M_{,y}} \tag{3.10}$$

Equation (3.8) gives

$$f_{ms} = a(\sin \theta \cos \varphi - \sin \varphi \cos \theta)$$

Using the trigonometric identity

$$\sin \omega = \frac{\tan \omega}{\sqrt{\tan^2 \omega + 1}}$$

and the condition

$$\frac{dM}{dm} = M_{,x} x_{,m} + M_{,y} y_{,m} = \frac{\partial m}{\partial m} = 1$$

we obtain for the interfringe spacing

$$f_{ms} = \left(M_{,x}^2 + M_{,y}^2\right)^{-1/2} \tag{3.11}$$

We can obtain equations analogous to Eq. (3.11) for the pitch of the specimen grating $S(x, y)$, p_s, and reference grating $R(x, y)$, p_r, as

$$p_s = \left(S_{,x}^2 + S_{,y}^2\right)^{-1/2} \tag{3.12a}$$

$$p_x = \left(R_{,x}^2 + R_{,y}^2\right)^{-1/2} \tag{3.12b}$$

Using Eq. (3.7), Eq. (3.11) takes the form

$$f_{ma} = \left[\left(S_{,x} + R_{,x}\right)^2 + \left(S_{,y} + R_{,y}\right)^2\right]^{-1/2} \tag{3.13}$$

for the additive pattern and the form

$$f_{ms} = \left[\left(S_{,x} - R_{,x}\right)^2 + \left(S_{,y} - R_{,y}\right)^2\right]^{-1/2} \tag{3.14}$$

for the subtractive moiré pattern.

The equation of the moiré boundary is obtained by

$$f_{ma} = f_{ms}$$

Substituting the values of f_{ma} and f_{ms} from Eqs. (3.13) and (3.14) we obtain the following equation for the commutation moiré boundary

$$\psi(x, y) = S_{,x} R_{,x} + S_{,y} R_{,y} = 0 \qquad (3.15)$$

The subtractive moiré is effective when $f_{ms} > f_{ma}$, or

$$\psi(x, y) = S_{,x} R_{,x} + S_{,y} R_{,y} > 0 \qquad (3.16)$$

From Eqs. (3.11) to (3.14) we obtain for the interfringe spacing f_{ms} and f_{ma} in terms of the pitches p_r and p_s of the two gratings

$$f_m = \frac{p_r p_s}{\left(p_r^2 + p_s^2 \pm 2\psi p_r^2 p_s^2\right)^{1/2}} \qquad (3.17)$$

where the positive and negative signs correspond to the additive and subtractive moiré patterns, respectively.

3.5 Relationships Between Line Grating and Moiré Fringes

Line gratings are mostly used in experimental mechanics. In this section, we will develop the relationships between the pitches and relative inclination of two lines gratings and the spacing and inclination of the resulting moiré fringes. The moiré fringes are straight lines that coincide with the diagonals of the rhombuses formed by the intersections of the lines of the two gratings. We will use the general method presented previously known as **geometric approach**.

Consider two lines gratings $S(x, y) = k$ and $R(x, y) = l$ of pitches, p and $p_1 = p(1 + \lambda)$, respectively, referred to a system of Cartesian coordinates Oxy (Fig. 3.7). The lines of the grating $S(x, y)$ are parallel to the x-axis, while the lines of the grating $R(x, y)$ make an angle θ with the x-axis. The line $k = 0$ of grating $S(x, y)$ and the line $l = 0$ of grating $R(x, y)$ pass through the origin O of the system Oxy.

The equations of the two gratings are expressed by

$$S(x, y) = k = \frac{y}{p} \qquad (3.18)$$

$$R(x, y) = 1 = \frac{y \cos \theta - x \sin \theta}{p(1 + \lambda)} \qquad (3.19)$$

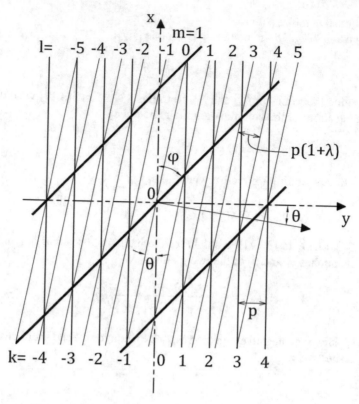

Fig. 3.7 Moiré fringes formed by two line gratings of different pitch p and $p(1 + \lambda)$ at an angle to each other

From Eq. (3.7) we obtain for the equation of the fringes of the subtractive moiré pattern

$$\frac{y}{p} - \frac{y \cos \theta - x \sin \theta}{p(1 + \lambda)} = m$$

or after rearranging the terms

$$y \left[\frac{\cos \theta - (1 + \lambda)}{\sin \theta} \right] + \frac{mp(1 + \lambda)}{\sin \theta} = x \tag{3.20}$$

The moiré fringes are straight lines with an angle of inclination φ with the x-axis and at a distance f (interfringe spacing) between them. Their equation can also be obtained from Eq. (3.19) by replacing θ with φ, and $p(1 + \lambda)$ with f. We obtain

$$y \cot \varphi - \frac{mf}{\sin \varphi} = x \tag{3.21}$$

By equating the y and the constant terms of Eqs. (3.20) and (3.21) we obtain

$$\cot \varphi = \frac{\cos \theta - (1 + \lambda)}{\sin \theta}, \frac{mf}{\sin \varphi} = -\frac{mp(1 + \lambda)}{\sin \theta} \tag{3.22}$$

From these equations, we obtain for the angle of inclination φ of the moiré fringes with respect to the x-axis

$$\sin \varphi = \frac{\sin \theta}{\sqrt{1 - 2(1 + \lambda) \cos \theta + (1 + \lambda)^2}} = \frac{p \sin \theta}{\sqrt{p^2 - 2pp_1 \cos \theta + p_1^2}} \tag{3.23}$$

and for the interfringe spacing f

$$f = \frac{p(1 + \lambda)}{\sqrt{1 - 2(1 + \lambda) \cos \theta + (1 + \lambda)^2}} = \frac{pp_1}{\sqrt{p^2 - 2pp_1 \cos \theta + p_1^2}} \tag{3.24}$$

Consider in Eqs. (3.23) and (3.24) that one grating with pitch p is the master grating and the other grating with pitch p_1 is the specimen grating. By measuring the angle φ and the interfringe spacing f, the angle θ and the p_1 are determined. For small angles, θ is related to the shear strain and p_1 to the normal strain. Even though this analysis is valid for a homogeneous field, it can be applied to a small area of a nonhomogeneous field for the determination of the strains.

For the special case of two inclined identical grating ($\lambda = 0, p_1 = p$) we obtain from Eqs. (3.23) and (3.24).

$$\sin \varphi = \cos(\theta/2), \ f = \frac{p}{2 \sin\left(\frac{\theta}{2}\right)}$$

The first equation indicates that $2\varphi = (\pi - \theta)$, which means that the moiré fringes bisect the supplementary angle $(2\pi - \theta)$ of angle θ. This result may also be derived from the condition that the moiré fringes are the shortest diagonals of the rhombuses formed by the intersections of the lines of the two gratings and bisect the angle $(2\pi - \theta)$.

For small angles θ (≈ 0) the first of the above equations indicate that $\varphi \approx (\pi/2)$, that is, the moiré fringes are perpendicular to the lines of the two gratings. From the second equation we obtain ($\sin (\theta/2) \approx \theta/2$).

$$\theta = \frac{p}{f} \tag{3.25}$$

In a deformable body, the angle θ represents the shear strain. In this respect, Eq. (3.25) shows that the shear strain is equal to the ratio of the pitch of the grating and the distance between two successive fringes. Similarly, Eq. (3.4) shows that the normal (Eulerian) strain ε is equal to the ratio of the pitch of the grating and

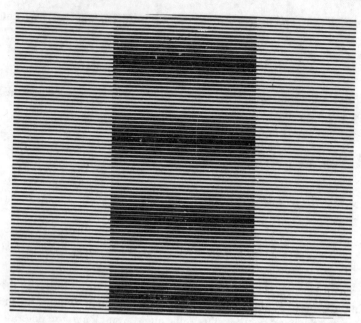

Fig. 3.8 Moiré fringes formed by two parallel gratings of slightly different pitch. The fringes are parallel to the lines of the gratings

the distance between two successive fringes. Both normal and shear strains are determined by the same Eq. (3.4) or Eq. (3.25), respectively.

For $\theta = 0$, that is when the lines of the two gratings are parallel, we obtain from Eqs. (3.23) and (3.24)

$$\varphi = 0, \ f = \frac{pp_1}{p_1 - p} = \frac{p(1 + \lambda)}{\lambda}$$

that is the moiré fringes are parallel to the lines of the gratings, and their distance is given by Eq. (3.5).

Figure 3.8 presents the moiré fringes formed by two parallel gratings of slightly different pitch. The fringes are parallel to the lines of the gratings and their distance is given by Eq. (3.5).

3.6 Moiré Patterns Formed by Circular, Radial and Line Gratings

Line gratings are mainly used in moiré analysis for measurement of rectilinear components of displacement. Circular and radial gratings may also be used for measurement of displacements in polar coordinates. In this section, we will study the moiré patterns formed by superposition of circular, radial, and line gratings by using the general equations of moiré fringes created by two families of curves $S(x, y) = k$ and $R(x, y) = l$ developed in Sect. 3.4. Superpositions of circular and line gratings create patterns of extreme beauty.

A **circular grating** consists of equispaced concentric circles. Consider two circular gratings of pitches p and $p(1 + \lambda)$ with their centers at distance $2c$ apart. Refer both gratings to a system of Cartesian coordinates with the x-axis passing through the centers of the circles of the gratings and its origin at the mid-distance between the centers of the gratings. The equations of the two circular gratings are given by

$$(x - c)^2 + y^2 = k^2 p^2 \quad (k = \pm 1, \pm 2, \ldots) \tag{3.26a}$$

$$(x + c)^2 + y^2 = l^2 p^2 (1 + \lambda)^2 \quad (l = \pm 1, \pm 2, \ldots) \tag{3.26b}$$

Using Eq. (3.7) we obtain the following equation for the moiré fringes obtained by the superposition of the two gratings

$$\{[(x + c)^2 + y^2] + [(x - c)^2 + y^2](1 + \lambda)^2 - m^2 p^2 (1 + \lambda)^2\}^2$$
$$= 4(1 + \lambda)^2 [(x + c)^2 + y^2][(x - c)^2 + y^2] \quad (m = \pm 1, \pm 2, \ldots) \tag{3.27}$$

The moiré pattern formed by two circular gratings of different pitches and at a distance apart for $\lambda > 0$ is shown in Fig. 3.9. Using Eq. (3.15) the commutation moiré boundary is obtained as

$$x^2 + y^2 = c^2 \tag{3.28}$$

which is the circumference of a circle of radius c. Inside the circle the additive moiré pattern is effective, while outside the circle the subtractive moiré pattern is effective.

When the two gratings are concentric $(c = 0)$, Eq. (3.27) becomes

$$x^2 + y^2 = \frac{m^2 p^2 (1 + \lambda)^2}{\lambda^2} \tag{3.29}$$

The moiré pattern is a family of concentric circles of radii $mp(1 + \lambda)/\lambda$.

Consider now the superposition of a circular grating of pitch p and a line grating of pitch $p(1 + \lambda)$. Take the origin of the coordinate system at the center of the circular grating and the lines on the line grating parallel to the x-axis. The circular and line gratings are expressed by

$$x^2 + y^2 = k^2 p^2, \, y = lp(1 + \lambda) \tag{3.30}$$

Using Eq. (3.7) we obtain the following equation for the moiré fringes

Fig. 3.9 Moiré pattern formed by a large displacement between two equispaced circular gratings of different pitch

$$(1 + \lambda)^2 x^2 + \lambda(2 + \lambda)y^2 \pm 2mp(1 + \lambda)y - p^2(1 + \lambda)^2 m^2 = 0 \qquad (3.31)$$

We can easily show that the commutation moiré boundary is the x-axis. The subtractive (minus sign in the above equation) moiré pattern is valid for $y > 0$, while the additive (plus sign in the above equation) moiré pattern is valid for $y < 0$. For $\lambda > 0$ the moiré fringes are ellipses, while for $\lambda < 0$ the moiré fringes are hyperbolas.

Consider now a **radial grating**. It is formed by a bundle of equiangular radial straight lines with variable pitch and line width equal to the interline width (Fig. 3.10). When the lines of the grating start at the origin of the coordinate system, the equation of the grating is

$$\text{arc } \tan(y/x) = l\,\alpha \qquad (3.32)$$

where α is the angle between two successive lines.

When a radial grating is superimposed with a line grating of equation $y = kp$, the equation of the moiré fringes is given by

$$\alpha y \pm p \text{ arc } \tan(y/x) = mp\alpha \qquad (3.33)$$

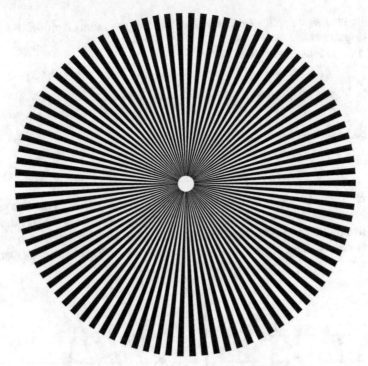

Fig. 3.10 A radial grating formed by a bundle of equiangular radial straight lines with variable pitch and line width equal to the interline width

It can easily be shown that the commutation moiré boundary is the y-axis. For x > 0 the subtractive moiré pattern (minus sign in Eq. (3.33)) is effective, while for x < 0 the additive moiré pattern (plus sign in Eq. (3.33)) is effective.

Consider now the superposition of two equiangular radial gratings with a distance 2c between their centers. Refer both gratings to a Cartesian system with origin at the mid-distance between the centers of the gratings. The equations of the grating are given by

$$\text{arc} \tan\left(\frac{y}{x+c}\right) = k\alpha \tag{3.34a}$$

$$\text{arc} \tan\left(\frac{y}{x-c}\right) = l\alpha \tag{3.34b}$$

Using Eq. (3.7) we obtain the following equation for the moiré fringes

$$x^2 \pm y^2 \pm 2xy \cot m\alpha = c^2 \quad m = \pm 1, \pm 2, \ldots \tag{3.35}$$

where the plus sign applies for the subtractive and the minus sign for the additive moiré, respectively.

The $\psi(x, y)$ function is given by

$$\psi(x, y) = \frac{x^2 + y^2 - c^2}{\alpha^2 \left[(x + c)^2 + y^2\right]\left[(x - c)^2 + y^2\right]} \tag{3.36}$$

The commutation moiré boundary is the circle

$$x^2 + y^2 = c^2 \tag{3.37}$$

This circle has its center at the mid-point between the centers of the two gratings and its circumference passes through the centers of the two gratings. Inside the circle, the additive moiré pattern is effective (Fig. 3.11). The moiré fringes are hyperbolas with their center at the origin. Outside the circle the subtractive moiré pattern is

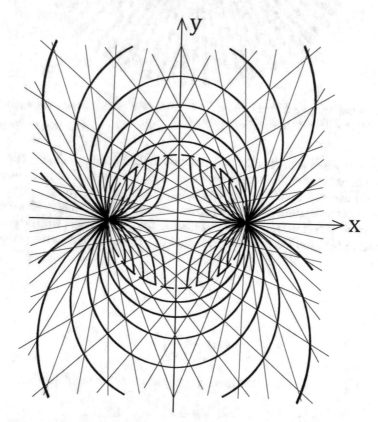

Fig. 3.11 Formation of moiré fringes by two radial gratings. The commutation moiré boundary is a circle passing through the centers of the gratings (shown with dotted lines)

effective. It consists of a family of circles having their centers on the y-axis and their circumference passes through the centers of the two radial gratings.

3.7 Measurement of In-Plane Displacements

So far we have studied the moiré patterns formed by superposition of two line, circular, or radial gratings. Particular attention was paid to line gratings of different pitches at an angle to each other. We established the equations between the pitches of the two gratings and their inclination angle on one hand and the angle of inclination and the distance between the resulting moiré fringes on the other hand.

We will now consider the general case of two line gratings one of which is attached to a two-dimensional specimen (Fig. 3.12). The specimen grating (SG) follows the deformation of the specimen, while the master grating (RG) is superposed to SG. The moiré fringes formed from the superposition of the gratings SG and RG are the shortest diagonals of the quadrangles formed by the gratings. Suppose that before deformation of the specimen the lines of the two gratings coincide. After deformation of the specimen, the points at which the lines of order $q - 1, q, q + 1$ of SG intersect the lines of order $q - 1, q, q + 1$ of RG do not move in a direction perpendicular to the lines of RG, and, therefore, belong to the same moiré fringe of zero order. For

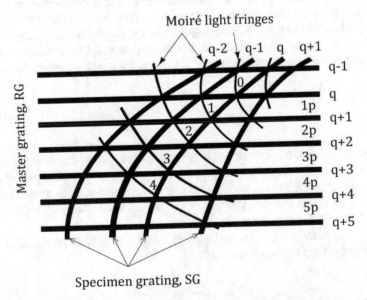

Fig. 3.12 Moiré fringes formed from the superposition of the specimen SG and master RG gratings are the shortest diagonals of the quadrangles formed by the gratings. They are the loci of points having the same value of the component of displacement in the direction perpendicular to the master grating

the same reason, the point at the intersection of line of order $q - 1$ of SG with the line of order q of RG has moved a distance equal to the pitch p of RG in a direction perpendicular to the lines of RG. This point belongs to the moiré fringe of first order. To the same fringe of first order belong the points at the intersection of the line q of SG with the line $q + 1$ of RG, the line $q + 1$ of SG with the line $q + 2$ of RG, etc. Similarly, the points at the intersections of the lines $q - 2, q - 1, q, q + 1$ of SG with the lines $q, q + 1, q + 2, q + 3$ of RG, respectively, belong to the same fringe of second order. These points have been moved a distance $2p$. The nth fringe is produced by a relative displacement of np.

Thus, *the moiré fringes are the loci of points with relative displacement in the direction perpendicular to the lines of the reference grating at the deformed state of the specimen equal to an integer number of the pitch of the master grating.* Each fringe is characterized by a parameter, n, which is called *fringe order*.

For the determination of the Cartesian components of the in-plane displacement two lines gratings are attached to the specimen before deformation with their principal directions parallel to the axes of a coordinate system. If the specimen has an axis of symmetry and the loads are symmetrical with respect to this axis, this axis is taken as an axis of the coordinate system. For the determination of the v-component of displacement along the y-axis, the lines of the specimen and reference gratings are parallel to the x-axis. The resulting moiré fringes from the superposition of the specimen and master gratings represent the contour lines of equal displacement v along the y-direction. Similarly, for the determination of the u-component of displacement along the x-axis the lines of the specimen and reference gratings are parallel to the y-axis. The resulting moiré fringes from the superposition of the specimen and reference gratings represent the contour lines of equal displacement u along the x-direction. The components of strain, $\varepsilon_{xx}, \varepsilon_{yy}, \gamma_{xy}$ for small deformations are given by

$$\varepsilon_x = \frac{\partial u}{\partial x}, \varepsilon_y = \frac{\partial v}{\partial x}, \gamma_{xy} = \frac{\partial u}{\partial y} + \frac{\partial v}{\partial x} \tag{3.38}$$

Instead of using two specimen gratings along two perpendicular directions we can use an orthogonally crossed grating and superimpose it twice with a master grating along its principal directions. We can also use two crossed gratings as specimen and reference gratings. In this case, two families of superposed moiré fringes are obtained representing u and v displacements.

As we mentioned previously, the Eulerian strain is defined from Eq. (3.3). Let us now define the Lagrangian strain ε_L as

$$\varepsilon^L = \frac{l_f - l_i}{l_i} = \frac{p_1 - p}{p} \tag{3.39}$$

where l_f is the final length after deformation and l_i is the initial length before deformation of fiber on the specimen. Note that the Lagrangian strain is referred to as the initial length, while the Eulerian strain is referred to the final length of the deformed fiber. The Lagrangian strain is related to the Eulerian strain ε^E by

$$\varepsilon^L = \frac{\varepsilon^E}{1 - \varepsilon^E} \tag{3.40}$$

When all three components of displacement u, υ and w along the x-, y-, and z-directions are known their derivatives along these directions ($\partial u/\partial x$, $\partial u/\partial y$, $\partial u/\partial z$, $\partial \upsilon/\partial x$, $\partial \upsilon/\partial y$, $\partial \upsilon/\partial z$, $\partial w/\partial x$, $\partial w/\partial y$, $\partial w/\partial z$) can be determined. For large deformations, the Lagrangian strains are determined from the following equations, where displacements must be differentiated along a line that is initially parallel to the coordinate system on the undeformed body and which has the initial dimensions of the body in that direction

$$\varepsilon_x^L = \left[1 + 2\frac{\partial u}{\partial x} + \left(\frac{\partial u}{\partial x}\right)^2 + \left(\frac{\partial \upsilon}{\partial x}\right)^2 + \left(\frac{\partial w}{\partial x}\right)^2 \right]^{1/2} - 1$$

$$\varepsilon_y^L = \left[1 + 2\frac{\partial \upsilon}{\partial y} + \left(\frac{\partial u}{\partial y}\right)^2 + \left(\frac{\partial \upsilon}{\partial y}\right)^2 + \left(\frac{\partial w}{\partial y}\right)^2 \right]^{1/2} - 1$$

$$\varepsilon_z^L = \left[1 + 2\frac{\partial w}{\partial z} + \left(\frac{\partial u}{\partial z}\right)^2 + \left(\frac{\partial \upsilon}{\partial z}\right)^2 + \left(\frac{\partial w}{\partial z}\right)^2 \right]^{1/2} - 1$$

$$\gamma_{xy}^L = \text{arc sin} \frac{\frac{\partial u}{\partial y} + \frac{\partial \upsilon}{\partial x} + \left(\frac{\partial u}{\partial x}\right)\left(\frac{\partial u}{\partial y}\right) + \left(\frac{\partial \upsilon}{\partial x}\right)\left(\frac{\partial \upsilon}{\partial y}\right) + \left(\frac{\partial w}{\partial x}\right)\left(\frac{\partial w}{\partial y}\right)}{\left(1 + \varepsilon_x^L\right)\left(1 + \varepsilon_y^L\right)}$$

$$\gamma_{yz}^L = \text{arc sin} \frac{\frac{\partial \upsilon}{\partial z} + \frac{\partial w}{\partial y} + \left(\frac{\partial u}{\partial y}\right)\left(\frac{\partial u}{\partial z}\right) + \left(\frac{\partial \upsilon}{\partial y}\right)\left(\frac{\partial \upsilon}{\partial z}\right) + \left(\frac{\partial w}{\partial y}\right)\left(\frac{\partial w}{\partial z}\right)}{\left(1 + \varepsilon_y^L\right)\left(1 + \varepsilon_z^L\right)}$$

$$\gamma_{zx}^L = \text{arc sin} \frac{\frac{\partial w}{\partial x} + \frac{\partial u}{\partial z} + \left(\frac{\partial u}{\partial x}\right)\left(\frac{\partial u}{\partial z}\right) + \left(\frac{\partial \upsilon}{\partial x}\right)\left(\frac{\partial \upsilon}{\partial z}\right) + \left(\frac{\partial w}{\partial x}\right)\left(\frac{\partial w}{\partial z}\right)}{\left(1 + \varepsilon_z^L\right)\left(1 + \varepsilon_x^L\right)} \tag{3.41}$$

The Eulerian strains are determined from the following equations, where the displacements must be differentiated along a line that is parallel to the coordinate direction on the deformed body and with the final dimensions of the fiber

$$\varepsilon_x^E = 1 - \left[1 - 2\frac{\partial u}{\partial x} + \left(\frac{\partial u}{\partial x}\right)^2 + \left(\frac{\partial \upsilon}{\partial x}\right)^2 + \left(\frac{\partial w}{\partial x}\right)^2 \right]^{1/2}$$

$$\varepsilon_y^E = 1 - \left[1 - 2\frac{\partial \upsilon}{\partial y} + \left(\frac{\partial u}{\partial y}\right)^2 + \left(\frac{\partial \upsilon}{\partial y}\right)^2 + \left(\frac{\partial w}{\partial y}\right)^2 \right]^{1/2}$$

$$\varepsilon_z^E = 1 - \left[1 - 2\frac{\partial w}{\partial z} + \left(\frac{\partial u}{\partial z}\right)^2 + \left(\frac{\partial \upsilon}{\partial z}\right)^2 + \left(\frac{\partial w}{\partial z}\right)^2 \right]^{1/2}$$

$$\gamma_{xy}^E = \text{arc } \sin \frac{\frac{\partial u}{\partial y} + \frac{\partial v}{\partial x} - \left(\frac{\partial u}{\partial x}\right)\left(\frac{\partial u}{\partial y}\right) - \left(\frac{\partial v}{\partial x}\right)\left(\frac{\partial v}{\partial y}\right) - \left(\frac{\partial w}{\partial x}\right)\left(\frac{\partial w}{\partial y}\right)}{\left(1 - \varepsilon_x^E\right)\left(1 - \varepsilon_y^E\right)}$$

$$\gamma_{yz}^E = \text{arc } \sin \frac{\frac{\partial v}{\partial z} + \frac{\partial w}{\partial y} - \left(\frac{\partial u}{\partial y}\right)\left(\frac{\partial u}{\partial z}\right) - \left(\frac{\partial v}{\partial y}\right)\left(\frac{\partial v}{\partial z}\right) - \left(\frac{\partial w}{\partial y}\right)\left(\frac{\partial w}{\partial z}\right)}{\left(1 - \varepsilon_y^E\right)\left(1 - \varepsilon_z^E\right)}$$

$$\gamma_{zx}^E = \text{arc } \sin \frac{\frac{\partial w}{\partial x} + \frac{\partial u}{\partial z} - \left(\frac{\partial u}{\partial z}\right)\left(\frac{\partial u}{\partial x}\right) - \left(\frac{\partial v}{\partial z}\right)\left(\frac{\partial v}{\partial x}\right) - \left(\frac{\partial w}{\partial z}\right)\left(\frac{\partial w}{\partial x}\right)}{\left(1 - \varepsilon_z^E\right)\left(1 - \varepsilon_x^E\right)} \tag{3.42}$$

Note that Eqs. (3.41) and (3.42) reduce to Eqs. (3.38) for small strains.

Figure 3.13 shows the moiré pattern of the displacement along the transversal direction (perpendicular to the applied load) of a perforated strip under tension. The reference and specimen gratings had the same pitch and their principal directions (perpendicular to the lines of the gratings) are oriented along the transversal direction. The fringes are the loci of points having the same value of transversal displacement. Figure 3.14 presents the moiré pattern of the vertical displacement field of a circular disk under diametral compression. The pitch of the grating is $p = 1/300$ in (0.085 mm)

Fig. 3.13 Moiré pattern of the transversal displacement field of a perforated plate in tension

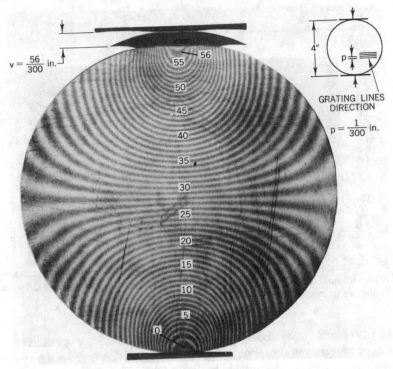

Fig. 3.14 Moiré pattern of the vertical displacement field of a circular disk under diametral compression. Fringes are the loci of points having the same vertical displacement. The specimen grating has a pitch $p = 1/300$ in. and its lines are horizontal

and its principal direction is oriented along the loading axis. The fringes are the loci of points having the same vertical displacement (along the loading direction). The fringe order at the top of the disk is 56, relatively to the bottom of the disk where it is zero and therefore, the displacement of the top relative to the bottom point of the disk is $56 \times (1/300) = 0.19$ in (4.74 mm).

The specimen (SG) and reference (RG) gratings usually have the same pitch. This is the case of large deformations. However, it may be advantageous for small deformations to use gratings with different pitches. If the pitch of the specimen grating SG is p and the pitch of master grating RG is $p(1 + \lambda)$ an initial fictitious deformation of λ is introduced. It is suitable for tensile fields to choose $\lambda < 0$, and for compressive fields $\lambda > 0$. Indeed, for a tensile field as the load has increased the difference between the pitches of the specimen and reference gratings increases, and therefore, the distance between the fringes decreases (Eq. (3.5)). In this way, dense moiré patterns are obtained. Similarly, for a compressive field more fringes are formed when we choose $\lambda > 0$. In calculating the strains the initial fictitious strain of λ should be subtracted.

The moiré fringes are formed by superposition of the reference and specimen gratings. Usually, the two gratings are put into direct contact. This can be avoided

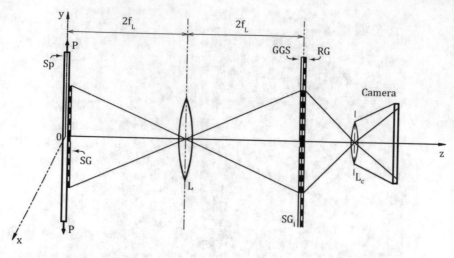

Fig. 3.15 Optical arrangement in the image moiré method. The image of the specimen grating is formed at unit magnification on a screen at some distance from the specimen where the reference grating is placed and superposed with the image of the specimen grating

by using the optical arrangement of Fig. 3.15. The image of the specimen grating is formed at unit magnification on a screen at some distance from the specimen where the reference (or master) grating is placed. The moiré pattern is formed by the mechanical interference of the two gratings on the screen. This optical arrangement allows different magnifications of the image of the specimen grating. Thus, it is possible to establish different values of the disparity term λ of the specimen grating of pitch p with respect to the reference grating of pitch $p(1 + \lambda)$. The contactless formation of moiré fringes is suitable, for example, high temperature experiments.

3.8 Measurement of Out-Of-Plane Displacements

The moiré methods studied in the previous sections of this chapter concerned with in-plane displacements u, υ and strains ε_x, ε_y, γ_{xy}. It was assumed that the out-of-plane displacements are small and they do not affect the in-plane displacements. In this section, we will present a method for measuring out-of-plane displacements independently from in-plane displacements. Determination of out-of-plane displacements is important in-plane stress problems because the out-of-plane strain is proportional to the sum of the two in-plane principal (or normal) stresses.

For the determination of out-of-plane displacements, we will present the **shadow moiré method**. The method uses the superposition of a grating and its shadow on the surface of the specimen. The surface of the specimen is coated with a matte finish and a master grating is placed in front of the surface (Fig. 3.16). The shadow of the master grating on the surface of the specimen constitutes the specimen grating. When

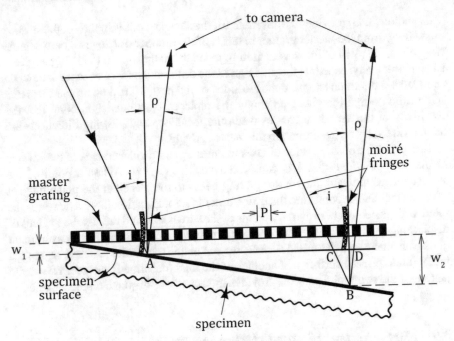

Fig. 3.16 Shadow moiré method for measuring out-of-plane displacements

a collimated beam of light is directed at an angle to the master grating interference of the master grating and its shadow on the surface takes place, and moiré fringes are formed.

Let i be the angle at which the collimated beam of light impinges on the specimen and ρ be the angle of viewing the obtained fringe pattern. Both angles are measured with respect to the normal to the master grating. Consider two adjacent fringe locations created from two points on the surface of the specimen at distances w_1 and w_2 from the master grating. The displacement of the "specimen grating" (the shadow of the master grating on the surface of the body) at direction perpendicular to the lines of the master grating, is

$$\delta = (CD) = (w_2 - w_1)(\tan\ i + \tan\ \rho)$$

For two points at successive moiré fringes, their relative displacement is equal to the pitch p of the master grating. Thus, we have

$$(w_2 - w_1)(\tan\ i + \tan\ \rho) = p$$

When n fringes are formed between the two points we have

$$\Delta w = w_2 - w_d = \frac{np}{\tan i + \tan \rho} \tag{3.43}$$

Equation (3.43) provides the difference of the distances of two points of the surface of the body from the master grating in terms of the number of fringes n between the two points, the pitch of the master grating p and the incidence and viewing angles, i and r, respectively. Note that the plus sign in the denominator applies when incidence and viewing directions are at opposite sides of the normal. If they are on the same side of the normal the plus sign should be replaced by minus sign. Note that the sensitivity of the method increases when the incidence and viewing directions are on opposite sides of the normal to the master grating.

In Eq. (3.43) a positive value of Δw indicates inward displacements of the surface of the body provided that n is considered positive ($n > 0$). When a point of the surface moves outward, the master grating needs to the placed at the plane tangent to the surface at that point, and the fringe order is zero. Equation (3.43) measures the relative displacement between two points of the surface. To find the absolute values of displacements, the displacement at some point needs to be known. For measurement of displacements of a deformed surface, we can obtain two moiré patterns, one of the initial (undeformed) surface and another of the final (deformed) surface. The moiré pattern of the two moiré patterns provides the actual out-of-plane displacements.

3.9 Measurement of Out-Of-Plane Slopes

Measurement of out-of-plane slopes of bending plates is important for the determination of stresses. The plate curvatures are obtained by simple differentiation of the slopes and stresses are linearly related to the curvatures. The plate curvatures ρ_x and ρ_y are related to the out-of-plane slopes by the following equations

$$\frac{1}{\rho_x} = -\frac{\partial^2 w}{\partial x^2}, \frac{1}{\rho_y} = -\frac{\partial^2 w}{\partial y^2} \tag{3.44}$$

and the normal stresses σ_x and σ_y are related to the curvatures by

$$\sigma_x = \frac{Ez}{1 - v^2}\left(\frac{1}{\rho_x} + v\frac{1}{\rho_y}\right), \sigma_y = \frac{Ez}{1 - v^2}\left(\frac{1}{\rho_y} + v\frac{1}{\rho_x}\right) \tag{3.45}$$

where E is the modulus of elasticity, v is the Poisson's ratio, and z is the coordinate variable along the thickness of the plate.

The out-of-plane displacement analysis presented in the previous section can be used for the determination of the stresses by twice differentiating the displacements. A direct determination of slopes allows the calculation of curvatures by performing one differentiation. A moiré method that gives directly the slopes $\partial w/\partial x$ and $\partial w/\partial y$ were developed by Ligtenberg.

The method is based on measurement of the displacement of the image of a master grating placed far from the plate after it is reflected from the surface of the plate. The

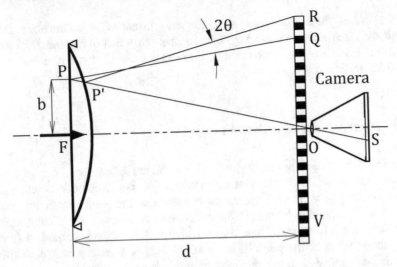

Fig. 3.17 Optical arrangement of the Ligtenberg method for measuring out-of-plane slopes

reflected image interferes with the master grating yielding fringes which represent the partial slopes of the plate. The specimen surface is made reflective instead of matte as in the previous case of measurement of out-of-plane displacements. The reflected image of the grating on the specimen does not depend on the angle of the incident light, and therefore, collimated light is not needed.

Figure 3.17 presents the optical arrangement of the **Ligtenberg method**. A coarse master grating is placed at a large distance d from the plate under study. The plate is viewed from a camera placed at a hole in the center of the grating. Consider a point P of the plate which is reflected at point Q on the screen. When the plate is deformed point P moves to point P' which is reflected to point R on the screen. Thus, loading of the plate creates a displacement of point Q to point R on the screen. By making a double exposure of the image of the master grating on the camera screen after its reflection form the unloaded and loaded plate a moiré pattern is formed. The distance QR for small deflections of the plate is given approximately by

$$QR \approx 2d\theta$$

where θ is the slope of the plate at point P.

Using the basic property of moiré fringes $QR = np$, we obtain

$$\theta = \frac{np}{2d} \tag{3.46}$$

where n is the fringe order.

Equation (3.46) gives the partial slope $\theta_x = \partial w/\partial x$ or $\theta_y = \partial w/\partial y$ of the surface of the plate depending on the orientation of the grating relative to the specimen. The distance between the specimen and the screen should be large to minimize the effect of out-of-plane displacements on the shift distance (QR). This effect is also minimized when a screen takes a cylindrical shape than that of a plane. By taking a cylindrical surface of the screen of diameter equal to $7d$ the error remains smaller than 0.3% for $b/d = 0.4$.

3.10 Sharpening of Moiré Fringes

According to the phenomenon of geometric moiré, the difference of displacements between two points on successive fringes is equal to the pitch of the grating. The density of the grating cannot exceed 40 lines/mm or the pitch cannot be smaller than 2.5×10^{-2} mm. For higher grating densities or smaller grating pitches, diffraction effects are introduced. Thus, the sensitivity of the moiré method is limited to the above value of the pitch of the grating. This is a serious drawback of the method compared with other methods of displacement measurement, e.g., electrical resistance strain gages. Attempts have been made to increase the sensitivity of moiré-fringe-sharpening and moiré-fringe-multiplication methods. Moiré-fringe-sharpening permits a more precise location of the position of the fringe and enhances the accuracy of the moiré method. In this section, we will present a method of moiré-fringe-sharpening.

According to Eq. (3.5) superposition of two parallel line gratings of pitches, p and $p(1 + \lambda)$ results in moiré fringes at distances $f \approx p/\lambda$. The light intensity is zero at the center of a dark fringe and increases linearly up to the adjacent bright fringe where becomes maximum. It then diminishes to zero at the next dark fringe. Figure 3.18a shows the triangular light intensity distribution along a line perpendicular to the moiré fringes formed by two intersecting line gratings of 50% transmittance (the width of the opaque bars of the gratings is equal to the width of the transparent slits). Minima or zero intensity occur at dark fringes and maxima of 50% of the intensity of the incident light occurs at bright fringes.

The above triangular intensity distribution can be altered by changing the ratio of the width of the opaque bars to the width of the transparent slits in both gratings. Consider the case where the one grating has wide opaque bars and narrow transparent slits, while the other grating has narrow opaque bars and wide transparent slits. Two gratings are called **complementary** when the ratio of the width of the opaque bars to the width of the transparent slits of one is reciprocal to the other. Figure 3.18b shows the light intensity distribution along a line perpendicular to the moiré fringes formed by two intersecting complementary line gratings. Note that the intensity pattern has a trapezoidal symmetric shape with minima of zero intensity at dark fringes and maxima of intensity 50% over a width $f(b - c)/p$, where b is the width of the opaque bars, c is the width of the transparent slits, p is the pitch and f is the interfringe spacing. The moiré fringes of Fig. 3.18b are sharper than those of Fig. 3.18a. Sharpening of moiré fringes can also be achieved with non-complimentary gratings. However, the difference of intensity between maxima and minima is reduced.

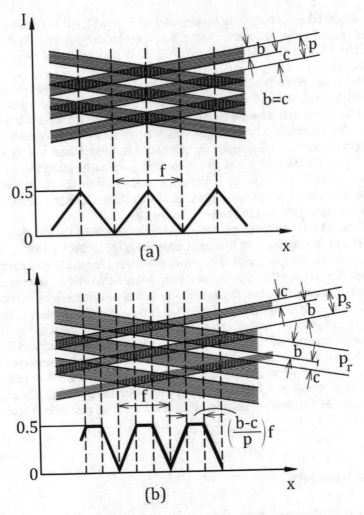

Fig. 3.18 Light intensity distribution along a line perpendicular to the moiré fringes formed by two intersecting line gratings of 50% transmittance (**a**) and by two complementary gratings (the ratio of the width of the opaque bars to the width of the transparent slits of one grating is reciprocal to the other) (**b**)

3.11 Moiré of Moiré

According to the moiré effect when two indexed families of curves $S(x,y) = k$ and $R(x, y) = l$ are superposed a third family $M(x, y) = m$ given by $k \pm l = m$ is generated (Eq. (3.7)). The two families of curves corresponding to plus and minus signs are called the additive and subtractive moiré patterns, respectively. The additive moiré pattern is not usually observed, while the subtractive moiré pattern dominates. *The moiré phenomenon results in subtracting quantities represented by two families of curves.* Moiré fringe patterns can be regarded as indexed families of curves. Thus, superposition of two moiré patterns results in a new moiré pattern called **moiré of moiré**. The moiré of moiré effect is used quite often in experimental mechanics. We refer two cases.

Consider the moiré fringes of u-displacements along the x-axis. Each fringe represents the loci of equal values of u. Obtain a copy of the moiré pattern of u-displacements and shift it by Δx along the x-direction. The two superposed families of curves form a new family of fringes which represent the partial derivatives of u with respect to x $(\partial u/\partial x)$. Similarly, we can obtain the derivatives of u with respect to y $(\partial u/\partial y)$ if we shift the moiré pattern of u-displacements along the y-axis, and take the moiré of the two patterns. In the same way, we obtain the derivatives $\partial v/\partial x$, $\partial v/\partial x$ of the v displacement along the x- and y-directions. Thus, *the moiré of moiré effect can be used in the differentiation of a function.*

As a second case, consider the interference fringe pattern formed by light rays reflected from the front and rear faces of a transparent specimen. Due to the variation of the thickness of the specimen (of the order of the wavelength of light) interference fringes are formed even though the specimen is unloaded. When the specimen is loaded an interference pattern is formed due to the variation of the stress-optical retardation of the light rays reflected from the two faces of the specimen. The interference pattern from the loaded and unloaded specimens presents the variation of the absolute stress-optical retardation of the specimen due to loading and the thickness of the specimen, respectively. When two such patterns are superposed with the specimen unloaded and loaded moiré fringes are obtained which indicate the isochromatics (loci of equal difference of principal stresses) and isopachics (loci of equal sum of the principal stresses) families of curves. This case of interference will be studied in Chap. 8.

Further Readings

1. Theocaris PS (1969) Moiré fringes in strain analysis. Pergamon Press
2. Durelli AJ, Parks VJ (1970) Moiré analysis of strain. Prentice-Hall Inc
3. Kafri O, Glatt I (1990) The physics of moiré metrology. Wiley-Interscience
4. Dally JW, Riley WF (1991) Experimental stress analysis. McGraw-Hill, pp 389–422
5. Parks VJ (1993) Geometric moiré. In: Kobayashi AS (ed) Handbook of experimental mechanics, 2nd edn. Society for Experimental Mechanics, pp 267–296
6. Patorski K, Kujawinska M (1993) Handbook of the moiré fringe technique. Elsevier
7. Post D, Han B, Ifju P (1995) High sensitivity moiré. Springer, pp 85–133
8. Cloud GL (1998) Optical methods in engineering analysis. Cambridge University Press, pp 174–203
9. Khan AS, Wang X (2001) Strain measurements and stress analysis. Prentice Hall, pp 30–93
10. Walker CA (ed) (2004) Handbook of moiré measurement. Institute of Physics Publishing
11. Amidror I (2007) The theory of the moiré phenomenon, vol. II Aperiodic layers, Springer
12. Han B, Post D (2008) Geometric moiré. In: Sharpe WN (ed) Handbook of experimental solid mechanics. Springer, pp 601–626
13. Amidror I (2009) The theory of the moiré phenomenon, vol I, 2nd edn. Periodic layers. Springer

14. Sciammarella CA, Sciammarella FM (2012) Experimental mechanics of solids. Wiley, pp 387–546
15. Shukla A, Dally JW (2014) Experimental solid mechanics, 2nd edn. College House Enterprises, pp 387–414

Chapter 4
Coherent Moiré and Moiré Interferometry

4.1 Introduction

Geometric moiré is a well-established full-field method for measuring displacements. A serious drawback of the method is its low sensitivity. The density of the gratings cannot exceed 40 lines/mm or the pitch cannot be smaller than 2.5×10^{-2} mm. The displacement difference between two successive fringes is equal to the pitch of the grating and this limits the sensitivity of the method. For higher density gratings diffraction effects enter which alter the nature of the geometric moiré effect. For measuring small displacements moiré methods with much higher sensitivity than geometric moiré have been developed. They use coherent illumination as opposed to geometric moiré which uses ordinary light.

In this chapter, we will present two high-sensitivity moiré methods, the method of **coherent or intermediate moiré** and the method of **moiré interferometry**. Unlike geometric moiré which is based on the phenomenon of obstruction of light, the two methods are based on wave optics and use the phenomena of diffraction by a grating, two-beam interference, and optical Fourier transformation by a lens.

Topics covered in this chapter include: diffraction by two superimposed parallel gratings, formation of patterns of the coherent or intermediate moiré method, optical filtering, fringe multiplication, and moiré interferometry.

4.2 Superposition of Two Diffraction Gratings

Consider two parallel sinusoidal amplitude or phase diffraction gratings, K and L illuminated by a monochromatic coherent beam at normal incidence (Fig. 4.1). Recall from Sect. 2.10.4 that a monochromatic collimated light beam after passing through a sinusoidal diffraction grating is divided into three beams: an undisturbed beam, called zero order, and two symmetrically deviated beams on either side of the zero order, called plus and minus first orders. The angle of deviation depends on the

E. E. Gdoutos, *Experimental Mechanics*, Solid Mechanics and Its Applications 269, https://doi.org/10.1007/978-3-030-89466-5_4

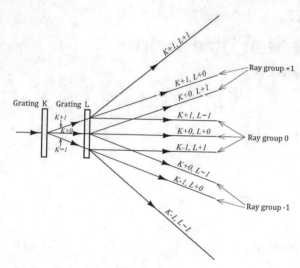

Fig. 4.1 Diffraction of a light ray by two parallel sine gratings with slightly different spatial frequencies

grating spatial frequency f and the wavelength of light λ. It is given by $\sin \theta \approx \theta = m\lambda f$ where $m = 0, +1, -1$ is the fringe order (Eq. (2.50)). When the two superposed gratings have the same pitch, their combination is equivalent to a single grating of the same pitch. The same set of diffraction orders appear as in a single grating.

Consider now the case of a small difference in the spatial frequencies of two gratings K and L, or the case when the spatial frequency of one grating is nearly an integral multiple of the other. The diffracted rays from the first grating are identified as $K + 0$, $K + 1$, and $K - 1$ for the zero, the first plus, and the first minus orders. Each of the three diffracted rays from the first grating is diffracted from the second grating producing three new rays. Thus, nine rays in total exit from the second grating. They are identified as $(K \pm J, L \pm J)$, $J = 0, 1$, where numbers of diffracted rays indicate orders at first and second gratings (e.g., $K + 1, L - 1$ means ray deviated into $+1$ order at grating K and then into -1 order at grating L). They are: $(K + 0, L + 1)$, $(K + 0 L + 0)$, $(K + 0, L - 1)$; $(K + 1, L + 1)$, $(K + 1, L + 0)$, $(K + 1, L - 1)$; $(K - 1, L + 1)$, $(K - 1, L + 0)$ $(K - 1, L - 1)$. These nine rays may be grouped as: $(K + 1, L + 1)$ (one ray); $(K - 1, L - 1)$ (one ray); $(K + 1, L - 1)$ and $(K + 0, L + 0)$ and $(K - 1, L + 1)$ named ray group #0 (three rays); $(K + 1, L + 0)$ and $(K + 0, L + 1)$ named group #1 (two rays), and $(K + 0, L - 1)$ and $(K - 1, L + 0)$ named ray group #-1 (two rays). The center group #0 of three rays is an attenuated version of the incident beam and does not need any further consideration. The single extreme rays $(K + 1, L + 1)$ and $(K - 1, L - 1)$ deviate and do not contribute to an optical effect. The intermediate two groups #$+1$ and #-1 contain two rays each, $(K + 1, L + 0)$, $(K + 0, L + 1)$ and $(K + 0, L - 1)$, $(K - 1, L + 0)$, one of which is diffracted from the first grating and the other is diffracted from the second grating. These two rays are nearly parallel because the pitches of the two

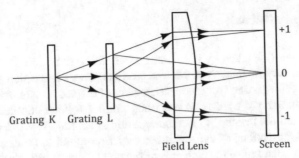

Fig. 4.2 Interference pattern formed by light diffracted from two sine gratings with slightly different spatial frequencies

gratings are nearly equal, or the pitch of one grating is nearly an integral multiple of the pitch of the other grating. Using a lens they interfere with one another (Fig. 4.2). The interference pattern depends on the angular difference of the two beams, which indicates the pitch difference of the two diffraction gratings. It is the moiré pattern of the two gratings. In conclusion, *the optical pattern obtained by two sinusoidal superimposed diffraction gratings illuminated by a coherent light beam is the moiré pattern of the two gratings.*

Thus, the end result of the superposition of two diffraction gratings including diffraction at each grating and interference of the resulting diffracted rays is surprisingly simple. *It is the moiré pattern of the two diffraction gratings.* This leads to the use of diffraction gratings for the determination of displacements.

The above analysis concerned sinusoidal amplitude diffraction gratings. For such gratings, three diffraction orders are obtained. For non-sinusoidal gratings, higher-order diffractions at each of the two gratings result. Thus, each group of nearly parallel rays exiting the second grating contains rays at different diffraction orders of the two gratings. Group of rays of higher diffraction orders corresponds to grating frequencies equal to the diffraction order times the fundamental basic grating spatial frequency. For example, consider the rays of the first diffraction order of a fine grating and the rays of the third diffraction order of a course grating. The pattern produced by these two groups of rays can be proved to be the same as the pattern produced by two gratings having nearly equal frequencies of three times the fundamental frequency of the course grating. The moiré fringes in the image correspond to the fringes that would be produced by two fine gratings. Thus, *the moiré fringes obtained by higher order diffraction rays correspond to grating frequencies equal to the basic frequency of the gratings times the diffraction order.* This is the basic concept for moiré fringe multiplication using diffraction gratings.

The above result is of great importance in practical applications. Instead of using two fine specimens and master gratings for moiré analysis we can use a course specimen grating that is easily applied and photographed and a fine master grating with multiple frequencies of the specimen grating and obtain the pattern formed by higher diffraction orders. The result is the same as if we would have used two fine gratings of the frequency of the master grating.

4.3 Moiré Patterns

The optical pattern obtained by illuminating two diffraction gratings by a coherent light beam is the moiré pattern of the two gratings. This observation leads the way of using diffraction gratings for the measurement of displacements. As in geometric moiré, two gratings are used. The specimen grating follows the deformations of the specimen and the master or reference grating is superimposed to the specimen grating. The deformations of the specimen vary from point to point and consist of translations and rotations of the elements of the body. Superposition of the specimen grating to the master grating leads to the creation of moiré fringes, in the same way as in geometric moiré. The moiré fringes for two-dimensional bodies represent u- and v-displacements along the x- and y-axes.

4.4 Optical Filtering and Fringe Multiplication

Diffraction gratings with coherent illumination offer many advantages in processing and manipulation of the optical patterns that cannot be obtained in geometric moiré. An optical processing system is illustrated schematically in Fig. 4.3. Two gratings, let say, a photograph of the specimen grating and a master grating is illuminated by a coherent collimated monochromatic light beam. A lens is placed behind the gratings. The lens acts as a Fourier optical processing system. At the focal plane of the lens a series of bright dots of the various groups of rays presented in the previous section appear. A screen with a stop or an aperture is placed at the transform plane. The aperture allows only rays of one group to enter the camera lens. In this way, a selection of a group of rays of different diffraction orders is made. All the other groups of rays are eliminated. Thus, moiré patterns of improved contrast and visibility are obtained.

The optical system of Fig. 4.3 can be used to provide moiré fringe sharpening and fringe multiplication. Note that if all ray groups are allowed to pass the diffraction plane the resultant pattern is identical to the moiré pattern produced by the two

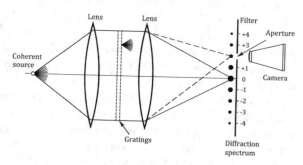

Fig. 4.3 An optical processing system for moiré fringe sharpening and fringe multiplication

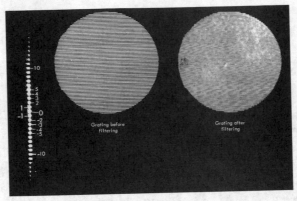

Fig. 4.4 Grating magnification by filtering all rays except those from the + 1 and − 1 diffraction orders

gratings. For the + 2, + 3, + 4, and + 5 diffraction orders multiplication by 2, 3, 4, and 5, respectively, occurs. Generally speaking, when light from diffraction order n is allowed to pass fringe multiplication by n is achieved. If rays from the one-plus dot (+ 1) and the one-minus (− 1) dot are allowed to pass the number of fringes is multiplied by a factor of two (2). If rays from the two-plus dot (+ 2) and the two-minus (− 2) dot are allowed to pass the number of fringes is multiplied by a factor of four (4), and so on. Fringe multiplication can be considered as multiplying the spatial frequency of the superimposed gratings by n. It increases the sensitivity and accuracy of moiré measurements. Multiplications by factors as high as 30 have been accomplished.

Figure 4.4 illustrates a grating magnification by filtering all rays except those from the + 1 and − 1 diffraction orders. The density of the obtained grating is twice that of the original grating. Furthermore, Fig. 4.5 illustrates the moiré fringe multiplication of a grating when all but one order of diffracted rays are filtered out by an aperture in the diffraction plane of an optical system like that shown in Fig. 4.3.

4.5 Advantages Offered by Coherent Moiré

Coherent or intermediate moiré is based on the phenomena of diffraction and interference of light. Since diffraction is an effect of interference, it can be said that coherent moiré is based on interference alone. It offers many advantages as compared to geometric moiré, as:

a. Increased sensitivity and improved fringe visibility.
b. Flexibility in processing optical patterns.
c. Spatial filtering using a lens as a Fourier transformer.
d. Fringe sharpening and multiplication of optical patterns.

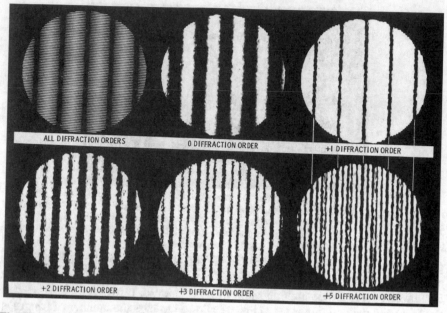

Fig. 4.5 Moiré fringe multiplication of a grating when all but one order of diffracted rays are filtered out by an aperture in the diffraction plane of an optical processing system

e. Use, of course, specimen gratings in combination with high-order diffraction groups, or with a master grating of multiple frequencies as that of the specimen grating.
f. The fringe data are collected for various loading states of the specimen, stored, and processed afterward. This allows elimination of any residual initial fringe patterns.
g. Reduction of optical noise through optical filtering.

4.6 Moiré Interferometry

4.6.1 Introduction

So far we have developed two moiré methods, the geometric moiré, and the coherent or intermediate moiré method. A third moiré method, the **moiré interferometry**, offers an upper end in sensitivity in measuring displacements. The method is capable to measure displacements of the order of the wavelength of light, that is, displacements of fractions of 1 μm. Moiré interferometry is useful in situations where high sensitivity is needed. In the following, we will present the optical arrangement of the method, two physical explanations of the obtained optical patterns and we will establish the equations for the determination of strains from the moiré fringes.

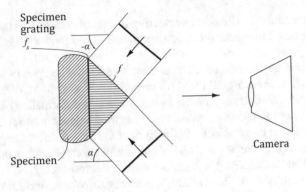

Fig. 4.6 Optical arrangement of moiré interferometry. The deformed specimen grating interacts with the virtual reference grating to form moiré fringes

4.6.2 Optical Arrangement

As in geometric and coherent moiré, a grating is placed on the specimen's surface. The spatial frequency of the grating is of the order of 1000–4000 lines/mm, whereas in geometric moiré it cannot exceed 40 lines/mm. This is the main practical difference between the two moiré methods. Another difference concerns the optical effects they are based on. Geometric moiré is based on the phenomenon of obstruction of light, whereas moiré interferometry, as coherent moiré, is based on the phenomena of diffraction and interference.

The specimen grating is produced from the interference of two plane waves of the same frequency and amplitude at an angle 2α between them (Fig. 4.6). The spatial frequency f_s (inverse of pitch) of the resulting grating is given by Eq. (2.39) as

$$f_s = \frac{2 \sin \alpha}{\lambda} \tag{4.1}$$

where λ is the wavelength of the light.

In this way, a very-high-frequency phase-type grating firmly bonded to the surface of the specimen is produced. For the placement of the grating on the specimen surface a replication technique is used. The grating is first produced on a photographic plate and is then transferred onto the specimen surface.

4.6.3 The Method

The specimen grating is illuminated by two coherent light beams at angles $+\alpha$ and $-\alpha$ (Fig. 4.6). The resulting optical effect can be explained as:

a. **First explanation**: The two intersecting beams of coherent light at angles $+\alpha$ and $-\alpha$ that illuminates the specimen interfere and produce a grating (virtual) at their plane of intersection with frequency given by Eq. (4.1). This virtual grating can be considered as a reference grating. The specimen grating and the virtual reference grating interfere and produce a moiré effect. When the specimen deforms the specimen grating follows the deformation of the specimen and produces moiré fringes which represent the contours of equal displacements in the direction perpendicular to the lines of the virtual reference grating, in the same way as in geometric moiré. Because the method is based on the phenomena of moiré and interference it is called **moiré interferometry**.

b. **Second explanation**: This explanation does not invoke the moiré effect, but only the phenomena of diffraction by a grating and interference. Under this explanation, the term moiré interferometry appears awkward and the method should be called **diffraction interferometry**. However, the term moiré interferometry dominated in the literature because of its relevance with the other two moiré methods, the geometric moiré, and the coherent moiré.

Each of the two incident light beams at angles $+\alpha$ and $-\alpha$ is diffracted by the specimen grating. The resulting two coherent diffracted beams in order to interfere must be normal to the specimen surface (as we will see below in order for this to occur the spatial frequency of the specimen grating must be half the spatial frequency of the virtual reference grating). Their angle of intersection is zero. They mutually interfere and form fringes, which represent the contours of equal displacements along the in-plane direction perpendicular to the lines of the specimen grating (Fig. 4.7).

An important result comes out from the second explanation. For the interference of the two diffracted beams from the specimen grating they must be normal to the specimen surface. The diffraction equation Eq. (2.51) applied to the specimen grating is

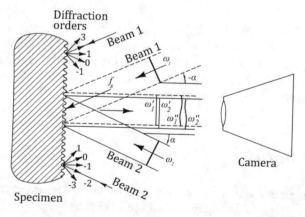

Fig. 4.7 Schematic diagram of the optical arrangement for moiré interferometry. The diffracted rays obtained by illuminating the specimen grating by two beams at angles $+\alpha$ and $-\alpha$ produce two wavefronts for the undeformed ($\omega_1{'}$ and $\omega_2{'}$) and ($\omega_1{''}$ and $\omega_2{''}$) for the deformed specimen.

$$\sin \beta_m = \sin \alpha + m\lambda f_s \qquad (4.2)$$

where β_m is the angle of the m-order diffraction ray, λ is the wavelength of light and f_s is the spatial frequency of the specimen grating.

Equation (4.2) for $\beta = 0$ and $m = +1$ (the angle of the first-order diffraction to be perpendicular to the specimen surface) becomes

$$\sin(-\alpha) + \lambda f_s = 0 \qquad (4.3)$$

The frequency f of the virtual reference grating produced by the light rays at an angle of incidence α with the specimen surface is

$$f = \frac{2 \sin \alpha}{\lambda}.$$

Introducing the value of $\sin\alpha$ into Eq. (4.2) we obtain

$$f_s = f/2 \qquad (4.4)$$

that is *the spatial frequency of* the *specimen grating must be half the spatial frequency of the virtual reference grating*. The same result is found for the -1 diffraction order.

4.6.4 Determination of Strains

Let us now consider that the specimen undergoes a uniform tensile stress ε_x. From Eq. (4.3) the frequency f_s of the specimen grating is

$$f_s = \frac{f/2}{1 + \varepsilon_x} \qquad (4.5)$$

Since f_s is not equal to $f/2$, the light from the first diffraction order of beam 1 does not emerge perpendicular to the grating, but at an angle β_1 with the perpendicular direction. The angle β_1 is given by the grating diffraction Eq. (4.2). By applying this equation for the specimen grating with

$$\sin \alpha = -\frac{\lambda f}{2}.$$

and using Eq. (4.5) we obtain

$$\sin \beta_1 \approx \beta_1 = \sin \alpha + \lambda f_s = -\frac{\lambda f}{2} + \lambda \frac{f/2}{1 + \varepsilon_x} = -\frac{\lambda f \varepsilon_x}{2} \qquad (4.6)$$

Similarly, the angle β_{-1} of the -1 diffraction order of beam 2 is given by

$$\beta_{-1} = \frac{\lambda f \varepsilon_x}{2} \tag{4.7}$$

The angular separation between these two coherent light beams is

$$\beta = \beta_1 - \beta_2 = \lambda f \varepsilon_x \tag{4.8}$$

The two light beams interfere and form an interference pattern of equidistant fringes. This is a grating with spatial frequency, F_{xx}, given by Eq. (4.1). We obtain for small angles β:

$$F_{xx} = \frac{2 \sin\left(\frac{\beta}{2}\right)}{\lambda} \approx \frac{\beta}{\lambda} = f \varepsilon_x \tag{4.9}$$

Equation (4.9) can be put in the form.

$$\varepsilon_x = \frac{p}{\delta} \tag{4.10}$$

where $\delta = 1/F_{xx}$ is the distance between two successive interference fringes and $p = 1/f$ is the pitch of the reference grating.

Equation (4.10) gives the strain ε_x as a ratio of the pitch p of the reference grating and the distance δ between two successive moiré fringes. It is the same equation as Eq. (3.4) of geometric moiré. Thus, *the strains in moiré interferometry and geometric moiré are given by the same equation.*

As a numerical example consider light from a Helium–Neon laser of wavelength $\lambda = 632.8 \times 10^{-9}$ m, a virtual reference grating of 2000 lines/mm and a normal strain of 0.2% (0.002 m/m). We obtain from the above equations:

$$\beta_1 = 0.00127 \text{ rad} = -0.073°$$
$$F_{xx} = 4 \text{ fringes /mm}$$
$$\delta = 0.25 \text{ mm}$$

Note that the fringe frequency F_{xx} is 0.2% of the frequency of the reference grating.

The above results can be extended to the general case of a nonhomogeneous two-dimensional displacement field. The obtained fringes represent the contours of the in-plane displacements u and v, in the same way as in geometric moiré. A major difference in geometric moiré and moiré interferometry is that in geometric moiré the reference grating is real, while in moiré interferometry the reference grating is imaginary consisting of a virtual image created by an interference pattern. Otherwise, the moiré fringes obtained in both methods represent the loci of equal displacements in directions perpendicular to the lines of the gratings.

Further Readings

1. Guild J (1956) The interference systems of crossed diffraction gratings. Oxford at the Clarendon Press
2. Durell AJ, Parks VJ (1970) Moiré analysis of strain. Prentice Hall, pp 35–54
3. Post D (1993) Moiré interferometry. In: Kobayashi AS (ed) Handbook of experimental mechanics, 2nd edn. Society for Experimental Mechanics, pp 267–296
4. Post D, Han B, Ifju P (1995) High sensitivity moiré. Springer, pp 135–417
5. Cloud GL (1998) Optical methods in engineering analysis. Cambridge University Press, pp 269–340
6. Post D, Han B (2008) Moiré interferometry. In: Sharpe WN (ed) Handbook of experimental solid mechanics. Springer, pp 627–653
7. Sciammarella CA, Sciammarella FM (2012) Experimental mechanics of solids. Wiley, pp 435–456
8. Shukla A, Dally JW (2014) Experimental solid mechanics, 2nd edn. College House Enterprises, pp 396–414

Chapter 5
Moiré Patterns Formed by Remote Gratings

5.1 Introduction

In this chapter, we present two moiré methods that use remote gratings for the determination of the gradients of the out-of-plane displacements or the sum of the two in-plane principal stresses for plane stress conditions. The moiré patterns are created by projecting the rulings of one grating onto the rulings of a second grating after interacting with the specimen. The first method developed by Theocaris and Koutsambessis [1–4] is based on geometric moiré and uses while light. The second method termed "Coherent Gradient Censor (CGS)" was developed by Rosakis and co-workers [5–17]. It is based on the diffraction of light by two gratings and uses coherent light. Both methods involve simple optical setups.

5.2 Geometric Moiré Methods

The experimental setup of a geometric moiré method is shown in Fig. 5.1 [5.1, 5.2]. Light rays emitted from a point source O become parallel through lens L_1. The light beam passes through the reference grating RG with principal direction along the ξ-axis (the rulings of the grating are parallel to the η-axis). A lens L_2 of focal length f_2 is placed at distance $l_2 = 2 f_2$ from the specimen. The image of grating RG is formed at unit magnification on a screen at a distance $2f_2$ from the lens L_2 (Eq. (2.7) of thin lenses). A grating MG with principal direction along the ξ-axis is placed on the screen. The superposition of the image of the grating RG and the grating MG forms a moiré pattern. When the pitch of the gratings RG is equal to the pitch of the grating MG a uniform (bright or dark) field is obtained. When the pitch of the grating RG is slightly different from the pitch of the grating MG parallel moiré fringes are formed (the distance between two successive fringes is dictated by Eq. (3.5)).

A transparent specimen Sp is placed near the grating RG. When the specimen has unloaded the image of the grating RG through the specimen Sp is focused on the

© The Author(s), under exclusive license to Springer Nature Switzerland AG 2022
E. E. Gdoutos, *Experimental Mechanics*, Solid Mechanics and Its Applications 269,
https://doi.org/10.1007/978-3-030-89466-5_5

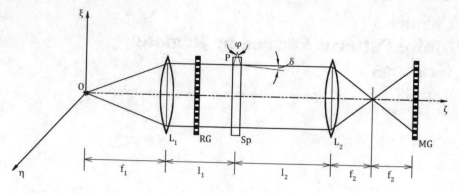

Fig. 5.1 Experimental arrangement of a geometric moiré method by projecting the rulings of one grating onto the rulings of a second grating after interacting with the specimen. The two gratings lie on parallel planes and their rulings are parallel

screen by the lens L_2 and is superposed with the grating MG. If the pitches of the two gratings RG and MG are equal and the surfaces of the specimen are completely parallel no optical effect is formed on the screen. However, when the surfaces of the specimen are not completely parallel an initial moiré pattern is formed which indicates the irregularities of the surfaces of the specimen. When the specimen has loaded the projection of the grating RG onto the screen through the specimen is distorted due to the changes of the relative angle between the two surfaces of the specimen. The moiré pattern created on the screen is directly related to the change of the angle between the surfaces of the specimen and, therefore, to the gradient of the two principal in-plane stresses in a direction perpendicular to the rulings of the grating.

We will derive a relation between the gradient of the sum of the principal stresses and the moiré pattern on the screen. We will ignore the change of the optical path of the light passing through the specimen due to loading. If φ the acute angle on the $O\xi\zeta$ plane between the deformed faces of the specimen at point P, the angle δ of deviation of the light ray after passing through the specimen, according to an approximate prism theory, is

$$\delta = (n - 1)\varphi \tag{5.1}$$

where n is the index of refraction of the material of the specimen.

The change Δh of the thickness h of the specimen under plane stress condition is given by

$$\Delta h = -\frac{\nu}{E}h(\sigma_1 + \sigma_2) \tag{5.2}$$

where ν is the Poisson's ratio, E is the modulus of elasticity, and σ_1 and σ_2 are the principal in-plane stresses.

The angle φ is given by

$$\varphi = \frac{\partial(\Delta h)}{\partial \xi} = -\frac{vh}{E}\frac{\partial(\sigma_1 + \sigma_2)}{\partial \xi} \tag{5.3}$$

From Eqs. (5.1) and (5.3) we obtain

$$\delta = -\frac{(n-1)vh}{E}\frac{\partial(\sigma_1 + \sigma_2)}{\partial \xi} \tag{5.4}$$

Equation (5.4) can also be obtained by considering the change of the optical path Δs of a light ray passing through the specimen. It is given by

$$\Delta s = (n - n_0)\Delta h + h\,\Delta n \tag{5.5}$$

where n is the index of refraction of the material of the specimen, n_0 is the index of refraction of the surrounding medium and h is the thickness of the specimen. Δ denotes the change of a quantity.

When the change of the refractive index of the specimen due to loading is small, the second term in the above equation can be omitted and Eq. (5.5) for $n = 1$ (air) becomes

$$\Delta s = (n-1)\Delta h \tag{5.6}$$

From Eqs. (5.6) and (5.2) we obtain

$$\Delta s = -(n-1)\frac{v}{E}h(\sigma_1 + \sigma_2) \tag{5.7}$$

The gradient of the optical path Δs, which is equal to the angle δ of the deviation of the light rays is obtained by differentiating the above equation as

$$\delta = \frac{\partial(\Delta s)}{\partial \xi} = -\frac{(n-1)vh}{E}\frac{\partial(\sigma_1 + \sigma_2)}{\partial \xi} \tag{5.8}$$

which is the same as Eq. (5.4).

Equations (5.4) or (5.8) show that the angle of deviation of a light ray or the gradient of the change of the optical path of the deformed specimen is proportional to the gradient along the ξ-axis of the sum of the principal stresses for conditions of plane stress. Thus, *the moiré fringes formed on the screen represent the gradients of the sum of the principal stresses*. If the grating RG is rotated by 90° we obtain the contours of the gradient along the η-axis.

Figure 5.2 shows the moiré fringes of loci of the gradient of the sum of the principal stresses along directions parallel and normal to the direction of the applied load of a strip with a circular hole under uniaxial tension for zero-initial fringe position (a) and for a non-zero-initial fringe position (b). Both patterns correspond to the same load.

(a)   **(b)**

Fig. 5.2 Moiré fringes of loci of the gradient of the sum of the principal stresses along directions parallel and normal to the direction of the applied load of a strip with a circular hole under uniaxial tension for zero initial fringe position (**a**) and for a non-zero initial fringe position (**b**). Both patterns correspond to the same load

Figures (5.3) and (5.4) show two other optical arrangements in which the obtained moiré fringes represent the gradients of the sum of the principal stresses. Figure 5.5 presents the moiré patterns of partial slopes along the principal axes of a perforated plate under uniaxial tension obtained by the experimental arrangement of Fig. (5.3). Part (a) of the figure corresponds to zero-initial fringe arrangement, and part (b) to a non-zero-initial fringe position obtained by receding the grating *RG* from the zero-initial fringe arrangement.

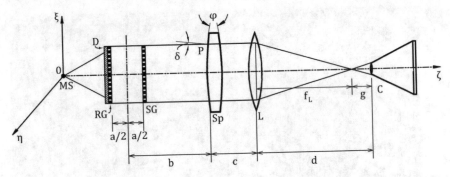

Fig. 5.3 Experimental setup of a geometric moiré method using two parallel gratings with parallel rulings. The moiré fringes are the loci of the gradient of the sum of the principal stresses along the principal directions of the gratings

5.3 The Coherent Grading Sensor (CGS) Method

5.3.1 Introduction

The method of coherent gradient sensor (CGS) is a shearing interferometric technique for measuring the gradient of the sum of the in-plane principal stresses for transparent materials or the gradient of the out-of-plane displacements for opaque materials. It is based on the effect of diffraction of light by two gratings. As it was shown in Sect. 4.2, the optical pattern obtained by two sinusoidal superimposed diffraction gratings illuminated by a coherent light beam is the moiré pattern of the two gratings. This pattern is modified by the deformations of the specimen and the resulting fringes represent the contours of the gradients of the in-plane principal stresses or the out-of-plane displacements.

5.3.2 Experimental Arrangement

The experimental arrangement of the transmission CGS method is shown in Fig. 5.6a. The specimen is illuminated by a collimated beam of coherent light. The transmitted beam is then incident normally on two diffractions gratings of the same pitch and parallel rulings placed on planes parallel to the specimen. A filtering lens L_1 is placed after the two gratings. The frequency content of the lens is displayed on its focal plane. Either of the ± 1 diffraction orders is blocked by a filtering aperture. The light passing through the filter is obtained on the image plane of the lens L_2. For opaque materials, the above optical arrangement is modified by using a beam splitter (Fig. 5.6b). The reflecting surface of the specimen is illuminated by a collimated beam of coherent light using a beam splitter. The reflected beam passes through the

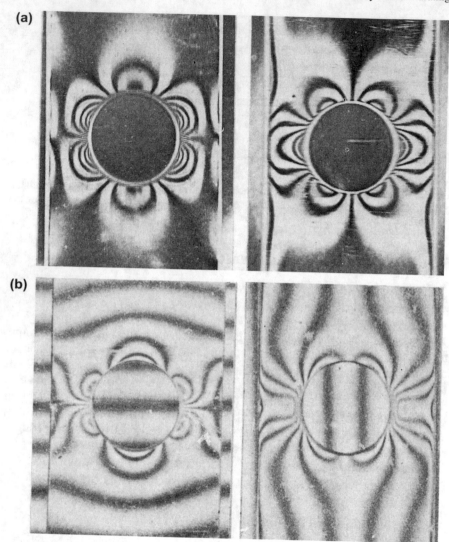

Fig. 5.4 Moiré fringes of loci of the gradient of the sum of the principal stresses along directions parallel and normal to the direction of the applied load of a strip with a circular hole under uniaxial tension for zero initial fringe position (**a**) and for a non-zero initial fringe position (**b**) for the experimental setup of Fig. 5.3

beam splitter, and as in the previous case, then passes through the two diffraction gratings, the lens, the filter plane, and is obtained on the image plane of a lens.

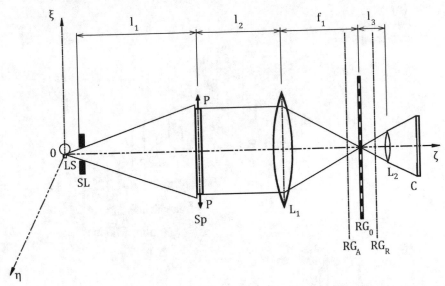

Fig. 5.5 Experimental setup of a geometric moiré method with two parallel gratings with parallel rulings. The moiré fringes are the loci of the gradient of the sum of the principal stresses along the principal directions of the gratings

5.3.3 Governing Equations

The equations of the image obtained in the above optical arrangement of the transmission or reflection CGS method will follow the developments of Sect. 4.2 using a simplified two-dimensional analysis [9]. For simplicity, it is assumed that the diffraction gratings have a sinusoidal transmission so that three diffraction rays are obtained at each grating.

First, consider that the specimen is undeformed and assume that the filtering lens blocks all but the -1 diffraction order. Following the analysis of Sect. 4.2, the diffracted light ray $r_{-1,0}$ (-1 diffraction order from the first grating, zero diffraction order from the second grating) deviates by an angle θ given by $\sin \theta \approx \theta = \lambda/p$, where p is the pitch of the grating and λ is the wavelength of light (Eq. (2.50)) (Fig. 5.7a). The deflected ray $r_{-1,0}$ exits the second grating at zero order with no deviation. The ray $r_{-1,0}$ makes an angle θ with the normal to the gratings. Consider now the ray $r_{0,-1}$ that exits at first grating at zero order with no deviation angle and the second grating at -1 order with deviation angle θ. Both angles are equal because the two gratings have the same pitch.

The rays $r_{-1,0}$ and $r_{0,-1}$ are parallel, and therefore, interfere. Let the magnitudes of the electric vectors of the two rays are given by (Eq. (2.25)).

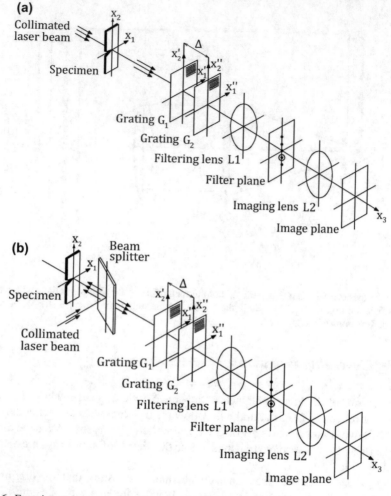

Fig. 5.6 Experimental arrangement of the transmission (**a**) and reflection (**b**) CGS method

$$E_1 = a_1 cos \frac{2\pi}{\lambda}(z_0 + \delta_1 - ct) = a_1 cos(\varphi_1 - \omega t)$$

$$E_2 = a_2 cos \frac{2\pi}{\lambda}(z_0 + \delta_2 - ct) = a_2 cos(\varphi_2 - \omega t) \qquad (5.9)$$

The intensity of the resulting interference pattern is (Eq. (2.27a))

$$I = a_1^2 + a_2^2 + 2a_1 a_2 cos(\varphi_2 - \varphi_1) \qquad (5.10)$$

where

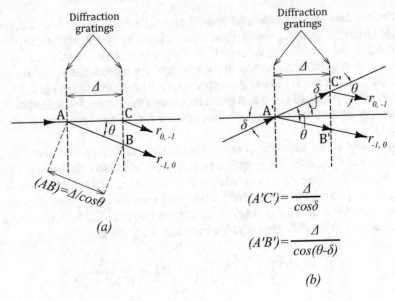

Fig. 5.7 Diffraction by two parallel grating for unloaded (**a**) and loaded specimen (**b**)

$$\varphi_2 - \varphi_1 = \frac{2\pi}{\lambda}(\delta_2 - \delta_1) \tag{5.11}$$

We have from Fig. 5.7a

$$\delta = \delta_2 - \delta_1 = \Delta\left[1 - (\cos\theta)^{-1}\right] = \Delta\left[1 - \left(1 - \frac{\theta^2}{2} + \cdots\right)^{-1}\right]$$

$$= \Delta\left[1 - \left(1 + \frac{\theta^2}{2} + \cdots\right)\right] = -\Delta\frac{\theta^2}{2} \tag{5.12}$$

where δ is the difference of the optical paths of the two interfering rays $r_{-1,0}$ and $r_{0,-1}$ and Δ is the distance between the two gratings (in deriving Eq. (5.12) we used the Taylor series expansion of the function $\cos x$ and $1/(1-x)$).

Constructive interference occurs when

$$\varphi_2 - \varphi_1 = \frac{2\pi}{\lambda}(\delta_2 - \delta_1) = \frac{2\pi}{\lambda}\Delta\frac{\theta^2}{2} = 2\pi N \tag{5.13}$$

where N is the fringe order.

Introducing the value of $\lambda = \theta\, p$ (from the diffraction equation) we obtain from Eq. (5.13)

$$\frac{\theta}{2} = \frac{Np}{\Delta} \tag{5.14}$$

Consider now that the specimen is deformed. The light rays passing through or reflected from the front face of the specimen undergo an optical path difference due to the deformation of the specimen. Thus, they impinge on the first grating at an angle φ with respect to the normal to the grating (or the specimen). The angle φ, as it was pointed out in the case of geometric moiré, is proportional to the gradient of the sum of the principal stresses for conditions of plane stress for transmitted rays or to the gradient of the thickness of the specimen for reflected rays, and is given by Eq. (5.4).

Let us now relate the angle φ to the interference pattern produced by the light rays $r_{-1,0}$ and $r_{0,-1}$ when the specimen is deformed. We follow the same procedure as in the previous case of the unloaded specimen. The only difference now is that the rays after passing through or deflected from the specimen impinge at an angle δ with the normal to the first grating. From the geometry of Fig. 5.7b we obtain the optical lengths of the diffracted rays between the two gratings

$$\delta_2' = (A'C') = \frac{\Delta}{\cos \delta'} \delta_1' = (A'B') = \frac{\Delta}{\cos(\theta - \delta)'}$$

and

$$\delta_2' - \delta_1' = \Delta \left(\frac{1}{\cos \delta} - \frac{1}{\cos(\theta - \delta)} \right) \approx \Delta \left(-\delta\theta + \frac{\theta^2}{2} \right) \tag{5.15}$$

The condition for constructive interference is

$$\frac{2\pi}{\lambda} (\delta_2' - \delta_1') = \frac{2\pi}{\lambda} \Delta \left(-\delta\varphi + \frac{\theta^2}{2} \right) = 2\pi N'$$

or

$$-\delta + \frac{\theta}{2} = \frac{N'\lambda}{\Delta}\frac{1}{\theta} = \frac{N'p}{\Delta} \tag{5.16}$$

where $p = \lambda/\theta$ according to the diffraction equation by a grating.

From Eqs. (5.14) and (5.16) we obtain

$$\delta = \frac{\theta}{2} - \frac{N'p}{\Delta} = \frac{(N - N')p}{\Delta} = \frac{np}{\Delta} \quad n = N - N' = 0, \pm 1, \pm 2, \ldots \tag{5.17}$$

From Eqs. (5.16) and (5.4) we obtain

$$-(n - 1)\frac{vh}{E}\frac{\partial(\sigma_1 + \sigma_2)}{\partial \xi} = \frac{np}{\Delta} \tag{5.18}$$

For transparent materials Eq. (5.18) becomes

$$ch\frac{\partial(\sigma_1 + \sigma_2)}{\partial\xi} = \frac{np}{\Delta} \tag{5.19}$$

where c is a stress-optical constant of the material of the specimen.

For the case of reflection from the front face of an opaque material, we obtain for the gradient of the out-of-plane displacement $w = \Delta h$

$$\delta = \frac{\partial(w)}{\partial\xi} = \frac{np}{\Delta} \tag{5.20}$$

For conditions of generalized plane stress we have

$$w = -\frac{vh}{E}(\sigma_1 + \sigma_2)$$

and Eq. (5.19) becomes

$$\frac{\partial(w)}{\partial\xi} = -\frac{vh}{E}\frac{\partial(\sigma_1 + \sigma_2)}{\partial\xi} = \frac{np}{\Delta} \tag{5.21}$$

Equations (5.19) and (5.21) for transparent and opaque materials, respectively, are the governing equation of the CGS method. They give the gradient of the sum of the principal stresses (transmission) or the gradient of the out-of-plane displacement (reflection) along the principal direction of a grating in terms of the pitch of the gratings (both gratings have the same pitch) and the distance between the gratings. Note that the sensitivity of the method increases by either increasing the distance Δ between the two gratings or by decreasing the pitch of the gratings.

Figure 5.8 presents the contours of the derivatives $(\partial u_3/\partial x_1)$ and $(\partial u_3/\partial x_2)$ of the out-of-plane displacement u_3 with respect to x_1 and x_2 for a three-point bend specimen made of AISI 4340 steel for (a) $P/P_0 = 0.38$ and (b) 0.61. P_0 is the plane stress limit load for the specimen and the axis x_1 is along the crack plane.

5.4 Comparison of the Geometric Moiré and the CGS Method

We presented two moiré methods that use remote gratings for the determination of the gradients of the out-of-plane displacements or the sum of the two in-plane principal stresses under conditions of plane stress for opaque and transparent materials, respectively. The methods present similarities and differences as:

Similarities

i.   Both methods use two remote gratings on parallel planes.

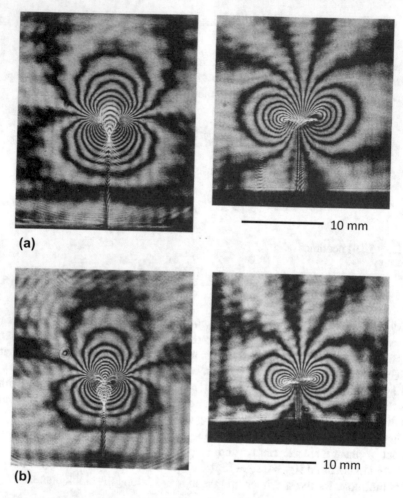

Fig. 5.8 Contours of $(\partial u_3/\partial x_1)$ and $(\partial u_3/\partial x_2)$ for a three-point bend specimen made of AISI 4340 steel **a** $P/P_0 = 0.38$ and **b** 0.61. P_0 is the plane stress limit load for the specimen [5]. Courtesy of Ares Rosakis

ii.   Both methods measure the gradient of the sum of the in-plane principal stresses for transparent materials (transmission mode) or the gradient of the out-of-plane displacement for opaque materials (reflection mode).
iii.   The optical arrangement for both methods is simple.
iv.   Both methods are relatively insensitive to vibrations.

Differences

i.   The Geometric moiré (GM) methods use amplitude gratings of frequency less than 40 lines/mm, while the CGS uses diffraction gratings of high frequencies.
ii.   GM uses noncoherent light, while CGS uses coherent light.

iii.   GM is based on the geometric moiré effect, while CGS is based on the effect of diffraction of coherent light by a grating.

iv.   The sensitivity of CGS is of the order of the wavelength of light, while the sensitivity of GM cannot exceed the smallest pitch of geometric moiré of 0.025 mm.

Further Readings

1. Theocaris PS (1969) Moiré fringes in strain analysis. Pergamon Press, pp 178–218
2. Theocaris PS, Koutsambessis A (1965) Slope measurement by means of moiré fringes. J Sci Instr 42:607–610
3. Theocaris PS, Koutsambessis A (1968) Surface topography by multisource moiré patterns. Exp Mech 8:82–87
4. Theocaris PS, Koutsambessis A (1968) The slit-source and grating method in plane stress problems. Strain 4:10–15
5. Tippur HV, Krishnaswamy S, Rosakis AJ (1991) A coherent gradient sensor for crack tip deformation measurements: analysis and experimental results. Int J Fract 48:193–204
6. Tippur HV, Krishnaswamy S, Rosakis AJ (1991) Optical mapping of crack tip deformations using the methods of transmission and reflection coherent gradient sensing—a study of crack tip K-dominance. Int J Fract 52:91–117
7. Tippur HV, Rosakis AJ (1991) Quasi-static and dynamic crack growth along bimaterial interfaces: a note on crack-tip field measurements using coherent gradient sensing. J Exp Mech 31:243–251
8. Krishnaswamy S, Tippur HV, Rosakis AJ (1992) Measurement of transient crack tip deformation fields using the method of coherent gradient sensing. J Mech Phys Sol 40:339–372
9. Tippur HV (1992) Coherent gradient sensing: a Fourier optics analysis and applications to fracture. Appl Opt 31:4428–4439
10. Bruck HA, Rosakis AJ (1992) On the sensitivity of coherent gradient sensing: Part I—a theoretical investigation of accuracy in fracture mechanics applications. Opt Lasers Eng 17:83–101
11. Mason JJ, Lambros J, Rosakis AJ (1992) The use of a coherent gradient sensor in dynamic mixed-mode fracture mechanics experiments. J Mech Phys Solids 40:641–661
12. Rosakis AJ (1993) Two optical techniques sensitive to gradients of optical path difference: the method of caustics and the coherent gradient sensor (CGS) In: Epstein JS (ed) VCH Publishers, pp 327–425
13. Bruck HA, Rosakis AJ (1993) On the sensitivity of coherent gradient sensing: Part II-an experimental investigation of accuracy in fracture mechanics applications. Opt Lasers Eng 18:25–51
14. Rosakis AJ (1993) Application of coherent gradient sensing (CGS) to the investigation of dynamic fracture problems. In: Shukla A, Guest (ed) Special issue of optics and lasers in engineering devoted to photomechanics applied to dynamic response of materials, vol 19, pp 3–41
15. Lee YJ, Lambros J, Rosakis AJ (1996) Analysis of coherent gradient sensing (CGS) by Fourier optics. Opt Lasers Eng 25:25–53
16. Rosakis AJ, Singh RP, Tsuji Y, Kolawa E, Moore NR Jr (1998) Full field measurements of curvature using coherent gradient sensing: application to thin film characterization. Thin Solid Films 325:42–54
17. Mello M, Hong S, Rosakis AJ (2009) Extension of the coherent gradient sensor (CGS) to the combined measurement of in-plane and out-of-plane displacement field gradients. Exp Mech 49:277–289

Chapter 6
The Method of Caustics

6.1 Introduction

The method of caustics is a simple optical method mainly used for the determination of stress intensity factors in crack problems under static and dynamic loading. Together with the method of geometric moiré, they are the only two optical methods of experimental mechanics included in this book that are based on geometric optics and not on the phenomena of interference or diffraction of light, as the other optical methods. Besides crack problems, the method can also be used to obtain optical patterns from reflecting surfaces.

In the following, we present the general equations of the method of caustics for reflecting surfaces and obtain the caustics created by axisymmetric ellipsoid mirrors. We apply the method for the determination of stress intensity factors in two-dimensional opening-mode and mixed-mode crack problems using optically isotropic or anisotropic materials.

6.2 General Equations for Reflecting Surfaces

Consider a reflecting surface referred to the system $Oxyz$ with equation

$$z = f(x, y) \qquad (6.1)$$

illuminated by a collimated light beam perpendicular to the plane Oxy (Fig. 6.1). When a reference screen is placed parallel to the plane Oxy at distance z_0 the deviation vector w of the reflected ray from a point $P(x,y)$ on the surface to a point $P'(x,y)$ on the screen, according to Snell's law of reflection, is given by

$$w = w_x i + w_y j \qquad (6.2)$$

© The Author(s), under exclusive license to Springer Nature Switzerland AG 2022
E. E. Gdoutos, *Experimental Mechanics*, Solid Mechanics and Its Applications 269,
https://doi.org/10.1007/978-3-030-89466-5_6

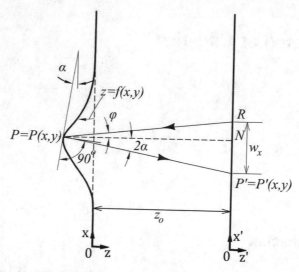

Fig. 6.1 Deviation of a reflected ray from point $P(x,y)$ on a surface $z = f(x,y)$ to point $P'(x,y)$ on a screen placed at a distance z_0 from the surface

with

$$w_x = (z - z_0)\tan 2\alpha, \quad w_y = (z - z_0)\tan 2\beta \tag{6.2a}$$

$$\tan\alpha = \frac{\partial f(x, y)}{\partial x}, \quad \tan\beta = \frac{\partial f(x, y)}{\partial y} \tag{6.2b}$$

$$\tan 2\alpha = \frac{2\partial f(x, y)/\partial x}{1 - \left[\frac{\partial f(x,y)}{\partial x}\right]^2}, \quad \tan 2\beta = \frac{2\partial f(x, y)/\partial y}{1 - \left[\frac{\partial f(x,y)}{\partial y}\right]^2} \tag{6.2c}$$

where i and j are the unit vectors referred to the projection $O'x'y'$ of the frame Oxy on the screen.

We refer vector w to the origin of the system $O'x'y'$. Then, the vector W which defines the position of the image point $P'(W_x, W_y)$ of point $P(x,y)$ of the surface on the plane $O'x'y'$ of the screen is given by

$$W = W_x i + W_y j = r + w \tag{6.3}$$

with

$$W_x = x + [f(x, y) - z_0]\frac{2\partial f(x, y)/\partial x}{1 - \left[\frac{\partial f(x,y)}{\partial x}\right]^2} \tag{6.3a}$$

$$W_y = y + [f(x, y) - z_0] \frac{2\partial f(x, y)/\partial y}{1 - \left[\frac{\partial f(x,y)}{\partial y}\right]^2} \qquad (6.3b)$$

Equations (6.3) maps the point $P(x,y)$ of the surface $z = f(x,y)$ to the point $P'(W_x, W_y)$ on the screen $O'x'y'$. The necessary and sufficient condition for points $P'(W_x, W_y)$ on the screen to belong to a curve is the zeroing of the Jacobian determinant of the transformation defined by Eqs. (6.3). This determinant is expressed as

$$J = \frac{\partial(W_x, W_y)}{\partial(x, y)} = \begin{vmatrix} \frac{\partial W_x}{\partial x} & \frac{\partial W_x}{\partial y} \\ \frac{\partial W_y}{\partial x} & \frac{\partial W_y}{\partial y} \end{vmatrix} = 0 \qquad (6.4)$$

Equation (6.4) defines a curve on the surface $z = f(x,y)$ called the **initial curve**. The system of Eqs. (6.3) and (6.4) defines a curve on the plane $O'x'y'$ of the screen called **caustic**. The caustic corresponds to the initial curve.

For every position of the screen defined by the distance z_0 from the plane Oxy, there is an initial curve on the surface and a corresponding caustic on the screen. By placing the screen at different distances z_0 from the plane Oxy we obtain on the screen a series of caustic curves which are the mappings on the screen of the corresponding initial curves on the surface.

If the slopes of the surface $z = f(x,y)$ are small, the squares of the derivatives of the function $f(x,y)$ in Eq. (6.3) can be neglected. Also, if the elevations of the surface $z = f(x,y)$ are small compared to z_0 (which is usually the case) the value of $f(x,y)$ can be neglected in Eq. (6.3). Under these conditions, Eq. (6.3) can be put in the form

$$W_x = x - 2z_0 \frac{\partial f(x, y)}{\partial x} \qquad (6.5a)$$

$$W_y = y - 2z_0 \frac{\partial f(x, y)}{\partial y} \qquad (6.5b)$$

Consider the case when the incident beam of light is not normal but subtends an angle φ with the normal to the plane Oxy at point $P(x,y)$. The angles 2α and 2β in Eq. (6.2a) are replaced by the angles $(2\alpha + \varphi)$ and $(2\beta + \varphi)$, respectively. Equation (6.2a) becomes

$$w_x = (z - z_0) \tan(2\alpha + \varphi), \quad w_y = (z - z_0) \tan(2\beta + \varphi) \qquad (6.6)$$

If the slopes and elevations of the surface $z = f(x,y)$ are small compared to z_0 ($\tan 2\alpha \approx 2 \tan\alpha$, $\tan \varphi \approx \varphi$), Eqs. (6.6) can be simplified to

$$w_x = -z_0(2 \tan\alpha + \tan\varphi), \quad w_y = -z_0(2 \tan\beta + \tan\varphi) \qquad (6.7)$$

Referring to the vector w $(w_x i + w_y j)$ of the origin of the system $O'x'y'$ we obtain for the vector W $(=W_x i + W_y j = r + w)$

$$W = W_x i + W_y j$$

with

$$W_x = x - z_0(2 \tan \alpha + \tan \varphi), \quad W_y = y - z_0(2 \tan \beta + \tan \varphi) \qquad (6.8)$$

Consider the case when the surface is illuminated by a point light source placed at a distance z_i from the plane Oxy along the z-axis. A magnification factor λ_m is defined by

$$\lambda_m = \frac{z_0 \pm z_i}{z_i} \qquad (6.9)$$

Equation (6.8) becomes

$$W_x = \lambda_m x - 2z_0 \frac{\partial f(x, y)}{\partial x}, \quad W_y = \lambda_m y - 2z_0 \frac{\partial f(x, y)}{\partial y} \qquad (6.10)$$

For an axisymmetric surface about the z-axis of the form $z = f(r)$, we obtain from Eqs. (6.6) for the magnitude $W(r)$ of the displacement vector referred to the origin O' of the system $O'x'y'$

$$W(r) = r + (z - z_0) \tan(2\alpha + \varphi) \qquad (6.11)$$

For small slopes and elevations of the surface $z = f(r)$, we obtain from Eq. (6.10)

$$W(r) = \lambda_m r - 2z_0 \frac{df(r)}{dr} \qquad (6.12)$$

Equation (6.4) renders the following equation for the initial curve of the caustic

$$\frac{dW(r)}{dr} = 0 \qquad (6.13)$$

From Eqs. (6.12) and (6.13) we obtain

$$\frac{d^2 f(r)}{dr^2} = \frac{\lambda_m}{2z_0} \qquad (6.14)$$

Equation (6.14) can be used for the determination of the curvature of the surface $z = f(r)$ at a point $r = r_0$. This point is defined by solving Eq. (6.14). For different values of the distance between the plate and the reference screen, z_0, the curvatures of the surface at different points $r = r_0$ are determined. Equation (6.14) applies for the determination of the curvatures and, therefore, the bending moments of axisymmetric plates in bending under small deflections.

6.3 The Ellipsoid Mirror

Consider an axisymmetric ellipsoid mirror with semiaxes a and b along the z and r axes, respectively, illuminated by a point light source S placed along its axis of symmetry z at a distance A from the r-axis (Fig. 6.2). A reference screen is placed at a distance z_0 from the r-axis. The equation of the surface $z = f(x,y)$ of the mirror referred to the system Orz is

$$z = f(x, y) = \frac{a}{b}\left(b^2 - r^2\right)^{1/2} \tag{6.15}$$

The equation of the caustic on the screen is given by Eq. (6.11) with

$$\tan \alpha = \frac{dz}{dr}, \quad \tan \varphi = \frac{r}{A + z} \tag{6.16}$$

Using Eq. (6.13) we obtain the following equation for the initial curve of the caustic

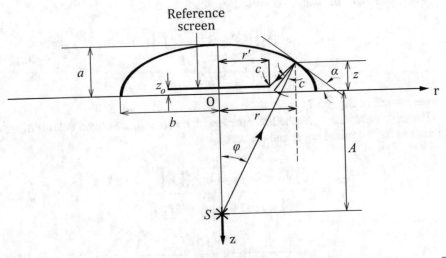

Fig. 6.2 Cross section of an axisymmetric ellipsoid mirror illuminated by a point light source S. The caustic is formed on the reference plane at distance z_0 from the Or axis

$$\frac{z_0}{b} = \frac{B_1\left[1 - \left(\frac{r}{b}\right)^2\right]^{1/2} + B_2}{\Delta_1\left[1 - \left(\frac{r}{b}\right)^2\right]^{1/2} + \Delta_2} \tag{6.17}$$

with

$$B_1 = 2\left(\frac{a}{b}\right)^2\left\{1 + \left[\left(\frac{a}{b}\right)^2 - 1\right]\left(\frac{r}{b}\right)^2 - \left(\frac{a}{b}\right)^2\right\}$$
$$+ \left\{1 - \left[\left(\frac{a}{b}\right)^2 + 1\right]\left(\frac{r}{b}\right)^2 - 2\left(\frac{a}{b}\right)^2\right\}\left(\frac{A}{b}\right)^2 \tag{6.17a}$$

$$B_2 = \left(\frac{A}{b}\right)\left(\frac{a}{b}\right)\left(3\left\{1 + \left[\left(\frac{a}{b}\right)^2 - 1\right]\left(\frac{r}{b}\right)^2\right\} - 4\left(\frac{a}{b}\right)^2\right) \tag{6.17b}$$

$$\Delta_1 = \left(\frac{A}{b}\right)\left\{1 + \left[\left(\frac{a}{b}\right)^2 - 1\right]\left(\frac{r}{b}\right)^2 - 4\left(\frac{a}{b}\right)^2\right\} \tag{6.17c}$$

$$\Delta_2 = \left(\frac{a}{b}\right)\left(\left\{1 + \left[\left(\frac{a}{b}\right)^2 - 1\right]\left(\frac{r}{b}\right)^2\right\}\right.$$
$$\left. - 2\left\{\left(\frac{a}{b}\right)^2 - \left[\left(\frac{a}{b}\right)^2 - 1\right]\left(\frac{r}{b}\right)^2\right\} - 2\left(\frac{A}{b}\right)^2\right) \tag{6.17d}$$

The equation of the caustic is given by

$$\frac{r'}{b} = \frac{2\left(\frac{a}{b}\right)\left\{\left[\left(\frac{a}{b}\right)^2 - 1\right] - \left(\frac{A}{b}\right)^2\right\}\left(\frac{r}{b}\right)^3}{\Delta_1\left[1 - \left(\frac{r}{b}\right)^2\right]^{1/2} + \Delta_2} \tag{6.18}$$

where r satisfies Eq. (6.17).

For the special case of a spherical mirror with $a = b = R$, we obtain the following equations for the initial curve and the caustic

$$\frac{z_0}{R} = \frac{\left(\frac{A}{R}\right)^2\left[1 + 2\left(\frac{r}{R}\right)^2\right]\left[1 - \left(\frac{r}{R}\right)^2\right]^{1/2} + \left(\frac{A}{R}\right)}{3\left(\frac{A}{R}\right)\left[1 - \left(\frac{r}{R}\right)^2\right]^{1/2} + 2\left(\frac{A}{R}\right)^2 + 1} \tag{6.19a}$$

$$\frac{r'}{R} = \frac{2\left(\frac{A}{R}\right)^2\left(\frac{r}{R}\right)^3}{3\left(\frac{A}{R}\right)\left[1 - \left(\frac{r}{R}\right)^2\right]^{1/2}2\left(\frac{A}{R}\right)^2 + 1} \tag{6.19b}$$

When the ellipsoid mirror is illuminated by a parallel beam of light ($A \to \infty$) we obtain from Eqs. (6.17) and (6.18)

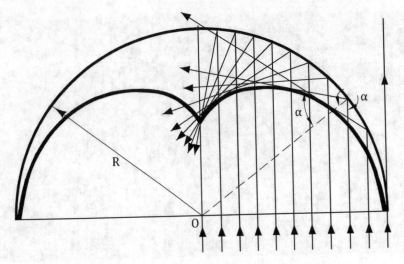

Fig. 6.3 Caustic formed by a spherical mirror illuminated by a parallel light beam

$$\frac{z_0}{b} = -\frac{\left\{1 - \left[\left(\frac{a}{b}\right)^2 + 1\right]\left(\frac{r}{b}\right)^2 - 2\left(\frac{a}{b}\right)^2\right\}\left[1 - \left(\frac{r}{R}\right)^2\right]^{1/2}}{2\left(\frac{a}{b}\right)} \tag{6.20a}$$

$$r'/b = (r/b)^3 \tag{6.20b}$$

For the case of a spherical mirror illuminated by a parallel beam of light, we obtain from Eqs. (6.19a, 6.19b) with $A \to \infty$ or from Eqs. (6.20a, 6.20b) with $a = b = R$

$$\frac{z_0}{R} = \left[\frac{1}{2} + \left(\frac{r}{R}\right)^2\right]\left[1 - \left(\frac{r}{R}\right)^2\right]^{1/2} \tag{6.21}$$

$$r'/R = (r/R)^3 \tag{6.22}$$

Figure 6.3 presents the caustic formed by a spherical mirror illuminated by a parallel light beam. The reflected rays from the mirror are obtained by applying Snell's law of reflection (the angle of reflection is equal to the angle of incidence). Note that the caustic is the envelope of the rays reflected from the mirror. The corresponding experimental caustic is shown in Fig. 6.4.

6.4 Intensity Distribution of Light Rays Reflected or Transmitted by a Transparent Specimen

Consider a light ray normally incident on a transparent specimen (Fig. 6.5). Part of the ray is reflected from the front face, while the other part passes through the thickness of the specimen. The transmitted ray meets the rear face of the specimen and part

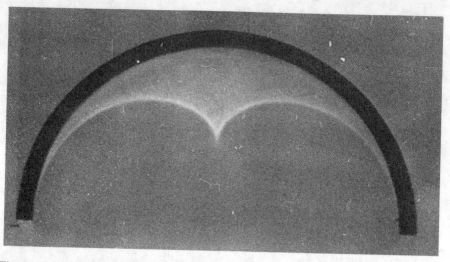

Fig. 6.4 Experimental caustic formed by a spherical mirror illuminated by a parallel light beam

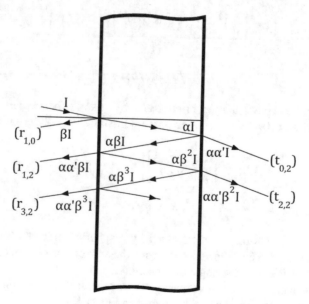

Fig. 6.5 Distribution of light intensity at various rays after successive reflections and refractions from the front and rear faces of a transparent plane specimen

of it is reflected, while the other part emerges from the rear face. The successive reflections from the two faces of the specimen are repeated so that an infinite number of rays emerge from the front and the rear faces of the specimen.

The intensity of light reflected from the front or the rear face of the specimen is the same at each reflection; let β be the reduction ratio of the intensity of reflected rays relative to the incident rays ($\beta < 1$). The reduction ratio for the rays that refract at the two faces differs for the rays that enter or emerge from the specimen. Let α and α' be the corresponding reduction ratios.

If I is the intensity of the incident light ray, the intensity of the first reflected ray is βI and the intensity of the first refracted ray is αI. This last ray emerges from the rear face of the specimen with intensity $\alpha\alpha'I$. At reflection from rear face, the ray passes through the thickness of the specimen with intensity $\alpha\beta I$. The intensity of light for the first three rays that emerge from the front face and the first two rays that emerge from the rear face of the specimen is shown in Fig. 6.5. According to the law of conservation of energy and assuming that the specimen does not absorb light energy we have

$$I = I_f + I_r \tag{6.23}$$

where I_f and I_r are the total intensity of light that emerges from the front and rear faces of the specimen, respectively.

We obtain for the intensities I_f and I_r

$$I_f = \beta I + \alpha\alpha'\beta\left(1 + \beta^2 + \beta^4 + \beta^6 + \cdots\right)I = \beta\left(1 + \frac{\alpha\alpha'}{1 - \beta^2}\right)I, \quad \beta < 1 \tag{6.24a}$$

$$I_r = \alpha\alpha'\left(1 + \beta^2 + \cdots\right)I = \frac{\alpha\alpha'I}{1 - \beta^2} \tag{6.24b}$$

Introducing the values of I_f and I_r into Eq. (6.23) we obtain

$$I = \beta\left(1 + \frac{\alpha\alpha'}{1 - \beta^2}\right)I + \frac{\alpha\alpha'I}{1 - \beta^2} \tag{6.25}$$

from which it follows that

$$\alpha\alpha' = (1 - \beta)^2 \tag{6.26}$$

We represent by $r_{k,l}$ ($r_{1,0}, r_{1,2}, r_{3,2}, \ldots$) the rays that emerge from the front face, where the indices k and l indicate the number of reflections and reflactions, respectively, and by $t_{k,l}$ ($t_{0,2}, t_{2,2}, t_{4,2}, \ldots$) the rays that emerge from the rear face of the specimen. We obtain for the intensities of these rays

$$I_{r,t_{k,l}} = \beta^k(1 - \beta)^l I \tag{6.27}$$

where $k = 1$, $l = 0, 2$ or $k = 3,5,7\ldots$, $l = 2$ for the rays that emerge from the front face and $k = 0, 2, 4\ldots$ and $l = 2$ for the rays that emerge from the rear face.

The coefficient β is given by [1]

$$\beta = \left(\frac{n-1}{n+1}\right)^2 \tag{6.28}$$

where n is the refractive index of the specimen.

For $n = 1.5$, which corresponds approximately to most common glasses and plastics, Eq. (6.28) yields $\beta = 0.04$. Then, we obtain from Eq. (6.27) the intensity of the rays that emerge from the front face of the specimen $I_{r_{1,0}} = 0.04I$, $I_{r_{1,2}} = 0.03686I$, $I_{r_{3,2}} = 0.00006\,I$, and for the intensity of the rays that emerge from the rear face of the specimen $I_{t_{0,2}} = 0.92160\,I$, $I_{t_{2,2}} = 0.00147I$. From these values of intensity, it is concluded that *only the rays* $r_{1,0}$, $r_{1,2}$ *and* $t_{0,2}$ *with intensities* $I_{r_{1,0}}$, $I_{r_{1,2}}$ *and* $I_{t_{0,2}}$ *must be considered* in optical effects of experimental mechanics related to transparent specimens.

6.5 Stress-Optical Equations

Consider a plane transparent specimen of a birefringent material. According to the **Neuman-Maxwell stress-optical law**, when a light beam is incident on the specimen it is divided into two linearly polarized beams along the principal stress directions and the index of refraction changes according to the following equation

$$\Delta n_1 = n_1 - n = b_1\varepsilon_1 + b_2(\varepsilon_2 + \varepsilon_3)$$
$$\Delta n_2 = n_2 - n = b_1\varepsilon_2 + b_2(\varepsilon_1 + \varepsilon_3) \tag{6.29}$$

where

n	index of refraction of the material of the specimen when it is unloaded
n_1, n_2	indices of refraction along the principal stress directions when the specimen is loaded
$\Delta n_1, \Delta n_2$	change of n_1, n_2
$\varepsilon_1, \varepsilon_2, \varepsilon_3$	principal strains
b_1, b_2	strain-optical constants of the material of the specimen

For linear elastic materials under conditions of plane stress ($\sigma_3 = 0$) we have according to Hooke's law

$$\varepsilon_1 = \frac{1}{E}(\sigma_1 - \nu\sigma_2), \quad \varepsilon_2 = \frac{1}{E}(\sigma_2 - \nu\sigma_1), \quad \varepsilon_3 = -\frac{\nu}{E}(\sigma_1 + \sigma_2) \tag{6.30}$$

where σ_1 and σ_2 are the principal stresses, and E and ν are the modulus of elasticity and Poisson's ratio.

When Eq. (6.29) is expressed in terms of stresses it takes the form

$$\Delta n_1 = n_1 - n = A\,\sigma_1 + B\,\sigma_2$$
$$\Delta n_2 = n_2 - n = B\,\sigma_1 + A\,\sigma_2 \qquad (6.31)$$

where

$$A = \frac{b_1 - 2vb_2}{E}, \quad B = \frac{b_2 - v(b_1 + b_2)}{E} \qquad (6.31a)$$

Consider a linearly polarized light beam normally traversing and passing through the specimen along the direction of the principal stress σ_1. Let A and B be two reference points along the light ray on opposite sides of the specimen (Fig. 6.6). The optical length (product of geometrical length times the index of refraction) between points A and B, $s_A = (AB)$, is given by

$$S_A = Ln_0 + d(n - n_0) \qquad (6.32)$$

where n_0 is the index of refraction of the medium that surrounds the specimen and d is the thickness of the specimen.

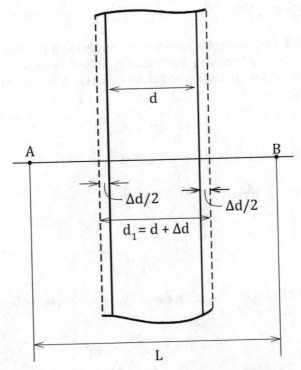

Fig. 6.6 An undeformed (continuous lines) and a deformed (dotted lines) specimen illuminated by a light ray passing through the specimen

When the specimen is loaded the optical path s_{T} between points A and B is

$$s_{T_1} = Ln_0 + (d + \Delta d)(n_1 - n_0)$$

where Δd denotes the change of the thickness of the specimen due to loading.

By omitting infinitesimal quantities of second order we obtain for the change of the optical length along the direction of the principal stress σ_1

$$\Delta s_{t_1} = s_{T_1} - s_A = d \, \Delta n_1 + (n_1 - n_0)\Delta d, \quad \Delta d = \varepsilon_3 \, d \tag{6.33}$$

Equation (6.33) can be written as

$$\Delta s_{t_1} = (a_t\sigma_1 + b_t\sigma_2)d \tag{6.34}$$

with

$$a_t = \frac{1}{E}[b_1 - 2vb_2 - v(n - n_0)] = A - \frac{v}{E}(n - n_0) \tag{6.34a}$$

$$b_t = \frac{1}{E}[b_2 - v(b_1 + (b_2)) - v(n - n_0)] = B - \frac{v}{E}(n - n_0) \tag{6.34b}$$

Equation (6.34) expresses the variation of the optical path of a light ray along the direction of the principal stress σ_1 due to loading of the specimen.

Similarly, we obtain for the variation of the optical path Δs_{t_2} along the direction of the principal stress σ_2

$$\Delta s_{t_2} = (b_t\sigma_1 + a_t\sigma_2)d \tag{6.35}$$

Equations (6.34) and (6.35) can be put in the form

$$\Delta s_{t_{1,2}} = c_t[(\sigma_1 + \sigma_2) \pm \xi_t(\sigma_1 - \sigma_2)]d \tag{6.36}$$

where

$$c_t = \frac{a_t + b_t}{2}, \quad \xi_t = \frac{a_t - b_t}{a_t + b_t} \tag{6.36a}$$

For the case of an optically inert material ($b_1 = b_2 = b$, $a_t = b_t = c_t$, $\xi_t = 0$) Eqs. (6.34) and (6.35) take the form

$$\Delta s_{t_1} = \Delta s_{t_2} = c_t(\sigma_1 + \sigma_2)d \tag{6.37}$$

where

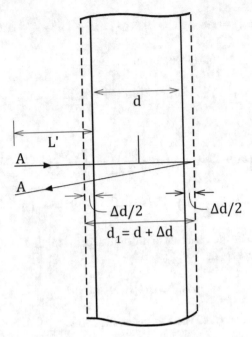

Fig. 6.7 An undeformed (continuous lines) and a deformed (dotted lines) specimen illuminated by a light ray reflected from the rear face of the deformed specimen

$$a_t = b_t = c_t = \frac{1}{E}[b(1-2v) - v(n-n_0)] \tag{6.37a}$$

Consider now the case when the incident light beam is reflected from the rear face of the specimen (Fig. 6.7). The optical path s_A for the light ray that starts from point A, traverses the specimen, is reflected from its rear face at point C, traverses the specimen again, and returns to point B ($B \equiv A$ since the light ray is perpendicular to the phase of the specimen) is given by

$$s_A = 2(L'n_0 + dn) \tag{6.38}$$

where L' is the length between point A and point D on the front face of the specimen when it is unloaded.

When the specimen is loaded the optical path length is

$$s_{T_1} = 2\left[\left(L' - \frac{\Delta d}{2}\right)n_0 + d_1 n_1\right] \tag{6.39}$$

From Eqs. (6.38) and (6.39) by omitting infinitesimal quantities of second order we obtain for the variation of the optical path length Δs_{r_1} of the light ray along the direction of the principal stress σ_1 which is reflected from the rear face of the specimen

$$\Delta s_{r_1} = s_{T_1} - s_A = 2\left[(n_1 - n)d + \left(n - \frac{n_0}{2}\right)\Delta d\right] \tag{6.40}$$

Using Eqs. (6.29) or (6.31) we obtain

$$\Delta s_{r_1} = 2(a_r\sigma_1 + b_r\sigma_2)d \tag{6.41}$$

where

$$a_r = \frac{1}{E}\left[b_1 - 2\nu b_2 - \nu\left(n - \frac{n_0}{2}\right)\right] = A - \frac{\nu}{E}\left(n - \frac{n_0}{2}\right) \tag{6.41a}$$

$$b_r = \frac{1}{E}\left[b_2 - \nu(b_1 + (b_2)) - \nu\left(n - \frac{n_0}{2}\right)\right] = B - \frac{\nu}{E}\left(n - \frac{n_0}{2}\right) \tag{6.41b}$$

Similarly, we obtain the variation of the optical path length Δs_{r_2} along the direction of the principal stress σ_2 as

$$\Delta s_{r_2} = 2(b_r\sigma_1 + a_r\sigma_2)d \tag{6.42}$$

Equations (6.41) and (6.42) can be put in the form

$$\Delta s_{r_{1,2}} = 2c_r[(\sigma_1 + \sigma_2) \pm \xi_r(\sigma_1 - \sigma_2)]d \tag{6.43}$$

where

$$c_r = \frac{a_r + b_r}{2}, \quad \xi_r = \frac{a_r - b_r}{a_r + b_r} \tag{6.43a}$$

For the case of an optically inert material ($b_1 = b_2 = b$, $a_r = b_r = c_r$, $\xi_t = 0$) Eqs. (6.41) and (6.43) take the form

$$\Delta s_{r_1} = \Delta s_{r_2} = 2c_r(\sigma_1 + \sigma_2)d \tag{6.44}$$

with

$$a_r = b_r = c_r = \frac{1}{E}\left[b(1 - 2\nu) - \nu\left(n - \frac{n_0}{2}\right)\right] \tag{6.44a}$$

6.6   Crack Problems

6.6.1   Introduction

In this section, we present the method of caustics for the determination of stress intensity factors in opening-mode and mixed-mode crack problems for opaque and optically isotropic materials. We also present the equations of caustics for optically anisotropic materials and discuss the three-dimensionality of the state of stress in the neighborhood of the crack tip.

6.6.2   Principle of the Method

In the method of caustics, a transparent or opaque specimen with a crack is illuminated by a light beam and the obtained optical effect is observed on a screen placed at some distance from the specimen. The reflected or transmitted rays undergo a change of their optical path due to the variation of the refractive index and the thickness of the specimen when it is loaded. The reflected or transmitted rays near the crack tip, due to the existing stress singularity, deviate and generate a highly illuminated three-dimensional surface in space (Fig. 6.8). When this surface is intersected by a

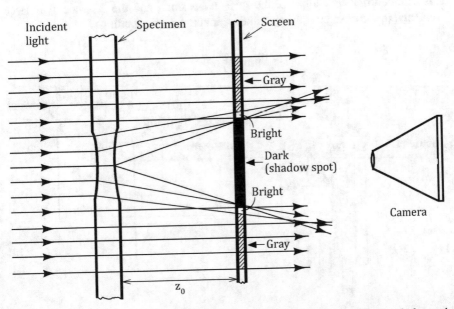

Fig. 6.8 Formation of a caustic curve by transmitted light rays from the neighborhood of a crack tip in a transparent specimen. The rays form a highly illuminated curve that surrounds the crack tip and separates the outside gray region from the inside dark region

reference screen, a highly illuminated curve, the so-called **caustic** is formed. The caustic surrounds the crack tip. It is the image on the reference screen of the initial curve on the specimen. The caustic separates an inside dark area from an outside gray area. The dark area is a result of the deflection of light rays. For transparent materials, three caustics are formed by the light rays reflected from the front and rear surfaces and those transmitted through the specimen. For opaque materials, one caustic is formed by the light rays reflected from the front surface of the specimen. The dimensions of the caustic are related to the state of stress in the neighborhood of the crack tip. The stress intensity factor, which governs the stress field near the crack tip, can be determined by measuring characteristic dimensions of the caustic, usually, its diameter perpendicular to the crack plane.

In the following we develop the equations of caustics for two-dimensional crack problems under conditions of generalized plane stress and relate the dimensions of the caustics to the stress intensity factors.

6.6.3 Opening-Mode Loading

Consider a transparent specimen illuminated by a parallel light beam and a screen at a distance z_0 downstream from the specimen placed parallel to the plane of the specimen (Fig. 6.9). A point P on the specimen is imaged at point P' on the screen. It is assumed that the slopes of the surface are small and the distance z_0 is large compared to the change of the specimen thickness due to loading.

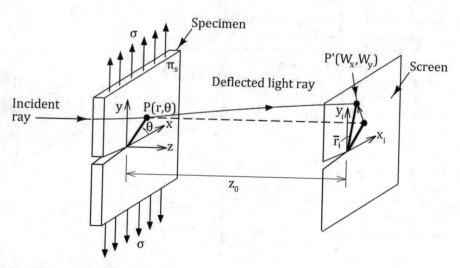

Fig. 6.9 A transparent specimen illuminated by a parallel light beam and a screen at a distance z_0 downstream from the specimen

For the case of a transparent specimen the mapping equation of point P to point P', according to Eq. (6.5), is given by

$$W_x = x - z_0 \frac{\partial(\Delta s)}{\partial x}$$
$$W_y = y - z_0 \frac{\partial(\Delta s)}{\partial y} \qquad (6.45)$$

where Δs is the change of the optical path of a ray passing through the specimen due to loading.

For an optically inert material, Δs is given by Eq. (6.37) for the light rays traversing the specimen and by Eq. (6.44) for the light rays reflected from the rear face of the transparent specimen. For the rays reflected from the front face of the specimen the stress-optical constant c_t in Eq. (6.37) should be replaced by the ratio v/E (v is the Poisson's ratio and E is the modulus of elasticity). For all three cases of light rays traversing the specimen, reflected from the rear face or from the front face of the specimen the transformation Eq. (6.45) applies with different values of the stress-optical constant c, which takes the values c_t, $2c_r$, or v/E, respectively.

The singular polar stresses at the tip of the crack for opening-mode loading governed by value of the stress intensity factor K_I are given by [2]

$$\sigma_r = \frac{K_I}{4\sqrt{2\pi r}}\left(5\cos\frac{\theta}{2} - \cos\frac{3\theta}{2}\right)$$
$$\sigma_\theta = \frac{K_I}{4\sqrt{2\pi r}}\left(3\cos\frac{\theta}{2} + \cos\frac{3\theta}{2}\right) \qquad (6.46)$$
$$\sigma_{r\theta} = \frac{K_I}{4\sqrt{2\pi r}}\left(\sin\frac{\theta}{2} + \sin\frac{3\theta}{2}\right)$$

where r is the distance from the origin of the coordinate system placed at the crack tip and θ is the polar angle.

From Eqs. (6.46) we obtain for the sum of the normal stresses σ_r and σ_θ

$$\sigma_r + \sigma_\theta = \frac{2K_I}{\sqrt{2\pi r}}\cos\frac{\theta}{2} \qquad (6.47)$$

The stress-optical path difference Δs using Eq. (6.37) is given by

$$\Delta s = c(\sigma_1 + \sigma_2)d = c(\sigma_r + \sigma_\theta)d = \frac{2cd K_I}{\sqrt{2\pi r}}\cos\frac{\theta}{2} \qquad (6.48)$$

where $c = c_t$ for the rays traversing the specimen, $c = 2c_r$ for the rays reflected from the rear face of the specimen, and $c = v/E$ for the rays reflected from the front face of the specimen.

From Eqs. (6.45) and (6.48) we obtain for the vector $W = W_x i + W_y j$ of the deviation of the light rays from point $P(r,\theta)$ on the specimen to point P' on the plane of the screen in polar coordinates ($x = r \cos\theta$, $y = r \sin\theta$)

$$W = W_x i + W_y j = \left(r \cos\theta + \zeta r^{-3/2} K_I \cos\frac{3\theta}{2} \right) i$$
$$+ \left(r \sin\theta + \zeta r^{-3/2} K_I \sin\frac{3\theta}{2} \right) j \tag{6.49}$$

with

$$\zeta = \frac{z_0 dc}{\sqrt{2\pi}} \tag{6.49a}$$

Equation (6.49) expresses the mapping of point $P(r,\theta)$ on the specimen to point $P'(W_x, W_y)$ on the screen. For a family of points, $P(r,\theta)$ a new family of points $P'(W_x, W_y)$ is generated. The pattern on the screen consists of a dark area and a gray area. The two areas are separated by a highly illuminated curve, the caustic. The condition for the existence of the caustic is the zeroing of the Jacobian determinant defined by Eq. (6.4) of the mapping dictated by Eq. (6.49). We obtain from Eqs. (6.4) and (6.49)

$$J = \begin{vmatrix} \frac{\partial W_x(r,\theta)}{\partial r} & \frac{\partial W_x(r,\theta)}{\partial\theta} \\ \frac{\partial W_y(r,\theta)}{\partial r} & \frac{\partial W_y(r,\theta)}{\partial\theta} \end{vmatrix}$$
$$= \begin{vmatrix} \cos\theta - \frac{3}{2}\zeta r^{-\frac{5}{2}} K_I \cos\frac{3\theta}{2} & -r\sin\theta - \frac{3}{2}\zeta r^{-\frac{3}{2}} K_I \sin\frac{3\theta}{2} \\ \sin\theta - \frac{3}{2}\zeta r^{-5/2} K_I \sin\frac{3\theta}{2} & r\cos\theta + \frac{3}{2}\zeta r^{-3/2} K_I \cos\frac{3\theta}{2} \end{vmatrix} = 0 \tag{6.50}$$

which yields

$$r = r_0 = \left(\frac{3K_I}{2\sqrt{2\pi}} z_0 dc \right)^{2/5} = (0.5984 z_0 dc K_I)^{2/5} \tag{6.51}$$

Equation (6.51) gives the equation of the points $P(x,y)$ on the specimen that corresponds to the points $P'(W_x, W_y)$ of the caustic on the image plane. The points $P(x,y)$ belong to a curve, called the **initial curve** of the caustic. We observe that the initial curve for the case of a crack subjected to opening-mode loading is a circle of radius $r = r_0$. Note that the size of the initial curve on the specimen varies with changing distance z_0 between the specimen and the image plane, the thickness of the specimen d, the stress-optical constant c, and the stress intensity factor K_I, which depends on the applied loads and the geometrical configuration of the plate. Thus, the initial curve of the caustic formed on the plane of the specimen depends on the geometry of the specimen, the dimensions of the optical arrangement, and the applied loads, and does not have a physical meaning. Its size varies with variation of the above quantities.

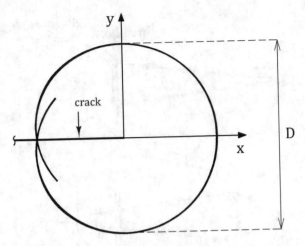

Fig. 6.10 Theoretical caustic formed by an optically isotropic material

Introducing the value of the initial curve into Eq. (6.49) we obtain the parametric equations of the caustic as

$$W_x = x' = r_0 \left(\cos \theta + \frac{3}{2} \cos \frac{3\theta}{2} \right)$$

$$W_y = y' = r_0 \left(\sin \theta + \frac{3}{2} \sin \frac{3\theta}{2} \right) \tag{6.52}$$

Equation (6.52) indicates that the caustic has the form of a generalized epicycloid (Fig. 6.10). From geometrical considerations, it can be shown that the diameter D of the caustic normal to the crack axis is related to the radius r_0 of the initial curve by

$$D = 3.16 r_0 \tag{6.53}$$

From Eqs. (6.51) and (6.53) we obtain

$$K_I = 0.0934 \frac{D^{2.5}}{z_0 c d} \tag{6.54}$$

Equation (6.54) *is used for the determination of stress intensity factor K_I by measuring the diameter D of the caustic at the crack tip which is perpendicular to the crack plane.* The quantities z_0, c, and d denote the distance of the reference screen from the specimen, the stress-optical constant of the material of the specimen, and the thickness of the specimen, respectively.

The above analysis corresponds to a parallel light beam. When the specimen is illuminated by a convergent or divergent light beam (Fig. 6.11) a magnification factor m of the optical arrangement should be introduced in Eq. (6.54). m is defined as the

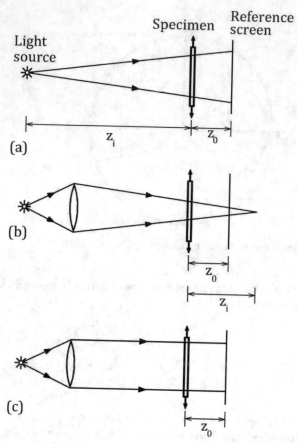

Fig. 6.11 Optical arrangement of a specimen illuminated by a divergent (a), convergent (b) or parallel light beam

ratio of a length on the reference screen where the caustic is formed divided by the corresponding length on the specimen. When the specimen is illuminated by a point light source placed at a distance z_i from the specimen m is given by

$$m = \frac{z_0 \mp z_i}{z_i} \tag{6.55}$$

where z_i is the distance of the impinging point light beam and the specimen. The plus (+) sign in the above equation applies for a divergent and the minus (−) sign for a convergent light beam.

By introducing the magnification number we obtain for K_I

$$K_I = 0.0934 \frac{D^{5/2}}{z_0 c d m^{3/2}} \tag{6.56}$$

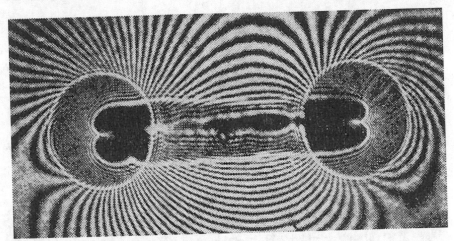

Fig. 6.12 Experimental caustics obtained at the tips of a central crack in a Plexiglas plate illuminated by light rays reflected from its front and rear faces. Two caustics are formed: the outer caustic corresponds to the light rays reflected from the rear face and the inner caustic to the light rays reflected from the front face of the specimen. Outside the outer caustics, the interference fringe pattern formed by the light rays reflected form the front and rear faces of the specimen is shown

Figure 6.12 presents an experimental caustic obtained in a Plexiglas plate with a central crack illuminated by light rays reflected from the rear face of the specimen. The interference fringe pattern outside the caustic is created by the light rays reflected from the front and rear faces of the specimen. The highly illuminated curve inside the caustic corresponds to a second caustic obtained from the light rays reflected from the front face of the specimen. The internal caustic is not a continuation of the external caustic because the optical constants for the two caustics are different (the stress-optical constant for the external caustic is c_r, while for the internal caustic is ν/E).

6.6.4 Mixed-Mode Loading

When the applied loads are not normal to the crack plane mixed-mode conditions apply in the vicinity of the crack tip. The singular stresses in polar coordinates are given in terms of the opening-mode K_I and sliding-mode K_{II} stress intensity factors by the following equations [2]

$$\sigma_r = \frac{K_I}{4\sqrt{2\pi r}}\left(5\cos\frac{\theta}{2} - \cos\frac{3\theta}{2}\right) + \frac{K_{II}}{4\sqrt{2\pi r}}\left(-5\sin\frac{\theta}{2} + 3\sin\frac{3\theta}{2}\right)$$

$$\sigma_\theta = \frac{K_I}{4\sqrt{2\pi r}}\left(3\cos\frac{\theta}{2} + \cos\frac{3\theta}{2}\right) + \frac{K_{II}}{4\sqrt{2\pi r}}\left(-3\sin\frac{\theta}{2} - 3\sin\frac{3\theta}{2}\right)$$

$$\sigma_{r\theta} = \frac{K_I}{4\sqrt{2\pi r}}\left(\sin\frac{\theta}{2} + \sin\frac{3\theta}{2}\right) + \frac{K_{II}}{4\sqrt{2\pi r}}\left(\cos\frac{\theta}{2} + 3\cos\frac{3\theta}{2}\right) \tag{6.57}$$

The sum of the stresses σ_r and σ_θ is given by

$$\sigma_r + \sigma_\theta = \frac{2K_I}{\sqrt{2\pi r}}\cos\frac{\theta}{2} - \frac{2K_{II}}{\sqrt{2\pi r}}\sin\frac{\theta}{2} \tag{6.58}$$

Working as in the previous case of opening-mode loading, we obtain the following equations of the caustic

$$W_x = x' = r_0\left[\cos\theta + \frac{2}{3}(1+\mu^2)^{-1/2}\cos\frac{3\theta}{2} - \frac{2}{3}\mu(1+\mu^2)^{-1/2}\sin\frac{3\theta}{2}\right]$$

$$W_y = y' = r_0\left[\sin\theta + \frac{2}{3}(1+\mu^2)^{-1/2}\sin\frac{3\theta}{2} + \frac{2}{3}\mu(1+\mu^2)^{-1/2}\cos\frac{3\theta}{2}\right] \tag{6.59}$$

with

$$r = r_0 = \left(\frac{3z_0dcK_I}{2\sqrt{2\pi}}\right)^{2/5}(1+\mu^2)^{1/5}, \quad \mu = K_{II}/K_I \tag{6.60}$$

The stress intensity factors K_I and K_{II} are obtained as

$$K_I = \frac{1.671}{z_0dcm^{3/2}}\left(\frac{D}{\delta}\right)^{5/2}\frac{1}{\sqrt{1+\mu^2}}, \quad K_{II} = \mu K_I \tag{6.61}$$

The ratio D/δ takes either of the values $r_0 = D_t/\delta_t$, D_l^{max}/δ_l^{max}, D_l^{min}/δ_l^{min}, where D_l^{max} and D_l^{min} are the maximum and minimum diameters of the caustic along the crack plane and D_t is the transverse (perpendicular to the crack plane) diameter of the caustic at the crack tip. The quantities δ_l^{max} and δ_l^{min} depend on the value of μ. For opening-mode loading $\delta_t = 3.16$, $\delta_l^{max} = \delta_l^{min} = 3$. Figure 6.13 presents the geometrical construction of the caustic for $\mu = 0.5$. In Fig. 6.14 four caustics corresponding to $\mu = 0, 0.25, 1.0$ and ∞ are shown. The values of $\mu = 0$ and ∞ refer to the cases of pure tension (opening-mode loading) and pure shear (sliding-mode loading), respectively. Note that the caustic for $\mu = 0$ is symmetric with respect to the x-axis, while the other caustics for $\mu \neq 0$ do not present a symmetry. Figure 6.15 presents the variation of the quantities $\delta_t = D_t/r_0$, $\delta^{max}=D_l^{max}/r_0$, $\delta^{min} = D_l^{min}/r_0$ versus μ. Note that for $\mu = 0$, $\delta_t = 3.16$, while for $\mu = 1$, $\delta_t = 2.99$. The variation of the ratio $(D_l^{max} - D_l^{min})/D_l^{max}$ versus μ is shown in Fig. 6.16.

For the experimental determination of stress intensity factors K_I and K_{II} the three diameters of the caustic D_t, D_l^{max}, D_l^{min} are measured. Then, the value of $\mu = K_{II}/K_I$ is determined from Fig. 6.16, and the value of δ (δ_t, δ_l^{max}, δ_l^{min}) from Fig. 6.15. Finally, having the values of μ and δ, the values of K_I and K_{II} are calculated from Eq. (6.61).

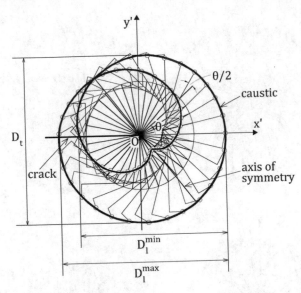

Fig. 6.13 Geometrical construction of a caustic for mixed-mode loading with $\mu = 0.5$

6.6.5 Anisotropic Materials

In optically anisotropic (birefringent) materials the change of the stress-optical path of a light ray is different along the directions of the two in-plane principal stresses σ_1 and σ_2. This change is given by Eq. (6.36) for the rays transmitted through the specimen and by Eq. (6.43) for the rays reflected from the rear face of the specimen. The plus (+) and minus (−) signs in these equations correspond to the two principal stresses. In such materials, two caustics are formed corresponding to the two principal stresses. Figure 6.17a, b presents the experimental double caustics at the crack tip formed by rays reflected from the rear face (a) and transmitted (b) through a specimen made of Polycarbonate of Bisphenol which is an optically anisotropic material. The specimen is loaded to opening mode.

Following the same procedure as in the case of optically isotropic materials presented previously, the opening-mode stress intensity factor K_I can be determined from Eq. (6.61) using the transverse (t) or longitudinal (l) diameter $D_{t,l}$ of the caustic with the corresponding values of $\delta_t^{\max}$ or $\delta_l^{\max}$ (instead of D and δ, with $\mu = 0$). The quantities $\delta_t^{\max}$ or $\delta_l^{\max}$ are functions of the anisotropy coefficient ξ of the material. Figure 6.18 presents the variation of $\delta_t^{\max}$ and $\delta_l^{\max}$ versus ξ for the transverse and longitudinal diameters of the caustic, respectively, for the outer and inner caustic.

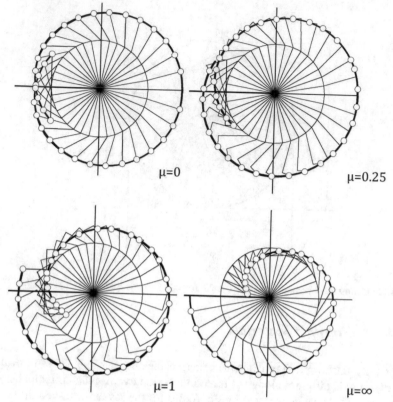

Fig. 6.14 Caustics corresponding to $\mu = 0$, 0.25, 10.0 and ∞

6.6.6 The State of Stress Near the Crack Tip

The previous analysis of the method of caustics was based on the assumption that plane stress conditions apply in the neighborhood of the crack tip. The value of the stress-optical constant c entered in Eqs. (6.55) and (6.60) for the determination of stress intensity factors K_I, K_{II} corresponds to plane stress conditions. However, the state of stress in the vicinity of the crack tip changes from plane strain very close to the tip to plane stress further away from the tip. Between these two regions, the stress field is three-dimensional. The value of c changes when the state of stress changes from plane strain to plane stress. In order to use the correct value of c for the determination of stress intensity factors K_I, K_{II} we should know the state of stress along the initial curve of the caustic.

It was established [3–10] that the state of stress becomes plane stress at distances r from the crack tip larger than approximately half the specimen thickness. Thus, *the radius of the initial curve of the caustic should be larger than half the specimen thickness d, $r_0 > d/2$*. If $r_0 < d/2$ the initial curve lies in the region where the state

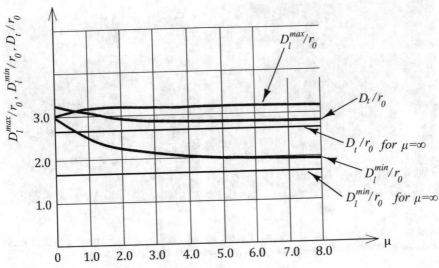

Fig. 6.15 Variation of the quantities $\delta_l^{max}=D_l^{max}/r_0$, $\delta_l^{min} = D_l^{min}/r_0$, $\delta_t = D_t/r_0$ versus μ

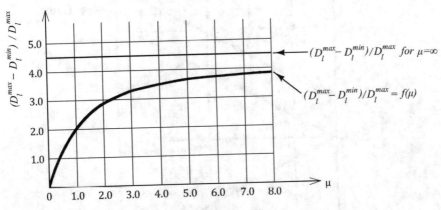

Fig. 6.16 Variation of the ratio $(D_l^{max} - D_l^{min})/D_l^{max}$ versus μ

of stress is three-dimensional and the plane stress value of c does not apply. For the correct application of the method of caustics, the condition $r_0 > d/2$ should be satisfied. The radius of the initial curve r_0 can be determined from Eq. (6.51) after the determination of the stress intensity factor and the above condition should be checked. If the condition $r_0 > d/2$ is not satisfied the experiment should be considered invalid and should be repeated by changing the values of the dimensions of the optical setup and the applied loads.

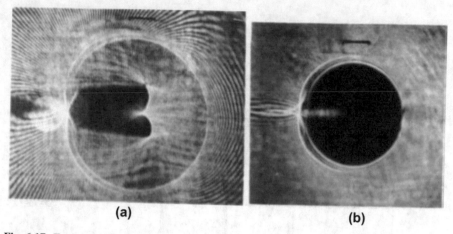

Fig. 6.17 Experimental double caustics formed by rays reflected from the rear face (**a**) and transmitted (**b**) through a birefringent specimen (courtesy of George Papadopoulos)

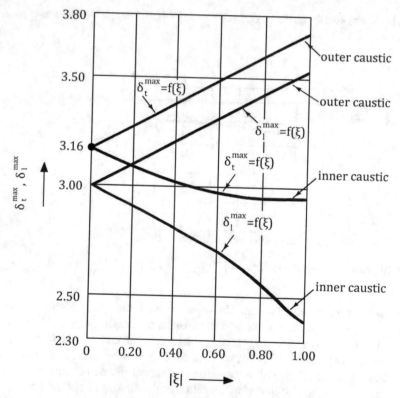

Fig. 6.18 Variation of δ_t^{max} and δ_l^{max} versus ξ for the outer and inner caustics

6.6.7 *Comparison of the Method of Caustics and Photoelasticity*

In the following we will compare the methods of caustics and photoelasticity for the determination of stress intensity factors in crack problems regarding the following criteria: the optical setup, the simplicity in the determination of stress intensity factors, the effect of crack tip radius, the effect of plate boundaries, the location of the crack tip and the changing state of stress near the crack tip.

6.6.7.1 Optical Setup

Photoelasticity is based on the phenomenon of interference of light, while the method of caustics is based on the laws of reflection and refraction of geometrical optics. The optical arrangement in both methods is simple. In photoelasticity, the specimen is inserted in the optical field of a plane or circular polariscope and monochromatic or white light is used. In the method of caustics, the specimen is illuminated by monochromatic or white light and the reflection or transmission image is received on a reference screen placed at some distance from the specimen.

The method of caustics can be used in a reflection or transmission arrangement. Actually, for a transparent material three caustics are obtained, two from the light rays reflected from the front and rear faces of the specimen, and one from the light rays transmitted through the specimen. As it was previously referred to, these three light rays carry most of the intensity of the infinite number of the light rays, that emerge from the front and rear faces of a transparent material. The intensity of the remaining light rays is insignificant. The method of photoelasticity is applied in transparent materials for transmission arrangements, while for opaque materials transparent birefringent (photoelastic) coatings are used. In both methods transmission arrangements record changes of thickness and refractive index, while in the method of caustics reflection arrangements from the front face of the specimen record only changes of thickness. In the method of photoelastic coatings reflection setups record changes of thickness and refractive index of the coating material.

6.6.7.2 Determination of Stress Intensity Factors

The value of stress intensity factor K_I in the method of caustics is determined by measuring the diameter of the caustic [Eqs. (6.56) and (6.61)]. This makes the determination of stress intensity factor very simple. Information for the evaluation of stress intensity factor is obtained from the initial curve surrounding the crack tip. The radius of the initial curve r_0 is determined from Eq. (6.51) for mode-I and Eq. (6.60) from mixed mode I and II by measuring the transverse diameter D_t of the caustic, the distance between the specimen and the reference screen where the caustic is formed

z_0, and the magnification m the optical setup. Thus, the region from which information is obtained is well known. Values of K_I can also be obtained by measuring other geometrical dimensions of caustic using simple evaluation formulas, analogous to Eqs. (6.56) and (6.61). K_I is determined either by the light rays transmitted through the specimen or by the light rays reflected from the rear or the front face of the specimen. Thus, transparent and opaque specimens can be studied.

In photoelasticity determination of K_I from isochromatic patterns is not an easy task. Near the crack tip, due to the existing singularity, there is concentration of isochromatic loops. Furthermore, the isochromatic pattern is influenced by nonlinear effects. To overcome this problem, data should be taken from the isochromatic patterns at critical distances away from the crack tip. This principle was used by Dally and coworkers [11, 12] who developed a technique to obtain K from data on isochromatics taken at distances away from the tip where the influence of nonlinearities and crack tip effects is minimal.

6.6.7.3   Effect of Crack Tip Radius

For the experimental determination of stress intensity factors in crack problems the crack is simulated by a slit. The stress field in the vicinity of the tip of a slit of radius ρ is given, according to Creager and Paris [13], by

$$\sigma_x = \frac{K_I}{(2\pi r)^{1/2}} \cos\frac{\theta}{2}\left(1 - \sin\frac{\theta}{2}\sin\frac{3\theta}{2}\right) - \frac{K_I}{(2\pi r)^{1/2}}\frac{\rho}{2r}\cos\frac{3\theta}{2} - \sigma_{0x}$$

$$\sigma_y = \frac{K_I}{(2\pi r)^{1/2}} \cos\frac{\theta}{2}\left(1 + \sin\frac{\theta}{2}\sin\frac{3\theta}{2}\right) + \frac{K_I}{(2\pi r)^{1/2}}\frac{\rho}{2r}\cos\frac{3\theta}{2}$$

$$\tau_{xy} = \frac{K_I}{(2\pi r)^{1/2}} \sin\frac{\theta}{2}\cos\frac{\theta}{2}\cos\frac{3\theta}{2} - \frac{K_I}{(2\pi r)^{1/2}}\frac{\rho}{2r}\sin\frac{3\theta}{2} \tag{6.62}$$

From the above equations, it is established that the sum of the normal stresses is independent of the radius ρ of the slit. For optically isotropic or opaque materials the caustic curve depends only on the sum of the principal stresses, and, therefore, it is independent of the radius ρ. On the contrary, the isochromatic fringes in photoelasticity depend on the difference of the principal stresses, and, therefore, depend on the radius ρ. This dependency distorts the isochromatic patterns in the vicinity of the crack tip and makes it mandatory to take measurements on the isochromatics at critical distances away from the tip of the slit that simulates the crack. However, these measurements should be taken in the K-dominance region where the stress field can be represented by the singular term only.

6.6.7.4   Effect of Plate Boundaries

The effect of plate boundaries enters in the equations of the stress field through the constant term σ_{0x}. The caustic depends on the gradient of the sum of the principal stresses, and, therefore, it is independent of σ_{0x}. On the other hand, in photoelasticity, the isochromatic fringes depend on the difference of the principal stresses, and, therefore, they are affected by the term σ_{0x}. The plate boundaries play a significant role in the form of the isochromatic loops and distort the direction of the loops either towards or away from the crack.

6.6.7.5   Location of Crack Tip

The exact location of the crack tip is necessary for the determination of crack speed in problems of running cracks. In photoelasticity, due to high concentration of isochromatics near the crack tip, location of the tip of the running crack is not an easy task. On the contrary, in the method of caustics, the exact location of the crack tip can easily be determined. This is achieved by locating the points at which the caustic surrounding the crack tip intersects the crack line on both sides of the crack tip. From the geometry of the caustic, it is deduced that the crack tip is at a distance 0.421D from the point of intersection of the caustic and the crack, and at a distance 0.527D from the point of intersection of the caustic and the crack ligament.

6.6.7.6   Effect of Changing State of Stress

The state of stress in the neighborhood of the crack tip changes from plane strain near the crack tip to plane stress at a critical distance away from the tip. The distance from the crack tip where plane stress prevails is approximately equal to half the specimen thickness (although, generally speaking, it depends on the crack length and the geometry of the cracked body). Between the plane strain and plane stress regions, the state of stress is three-dimensional. In the optical method of caustics, plane stress conditions are assumed to dominate near the crack tip. This implies that the initial curve which is mapped onto the caustic should lie in the plane stress region. The stress-optical constant changes significantly from plane strain to plane stress. Thus for the correct determination of stress intensity factors, the initial curve should lie in the plane stress region in order to use the value of the stress-optical constant for conditions of plane stress.

In photoelasticity, the isochromatic fringes depend on the difference of the optical path along the two principal stress directions. This difference for the light rays either traversing the specimen or reflected from the rear face of the specimen is independent of the state of stress and is the same for plane stress or plane strain conditions [2, pp 419–420]. This constitutes a great advantage of the method of photoelasticity over the method of caustics.

Further Readings

1. Manogg P (1964) Anwendung der Schattenoptic zur Untersuchung des Zerreissvorgangs von Platten, Dissertation, Freiburg, Germany
2. Gdoutos EE (2020) Fracture mechanics—an Introduction, 3rd edn. Springer
3. Levi N, Marcal PV, Rice JR (1971) Progress in three-dimensional elastic-plastic stress analysis for fracture mechanics. Nucl Eng Des 17:64–75
4. Rosakis AJ, Ravi-Chandar K (1986) On crack tip Stress State - An experimental evaluation of three-dimensional effects. Int J Sol Struct 22:121–134
5. EI M, Huang W, Gdoutos EE (1991) A Study of the three-dimensional region at crack tips by the method of caustics Eng Fract Mech 39:875–885
6. Konsta-Gdoutos M, Gdoutos EE, Meletis EI (1992) The state of stress at a crack tip studied by caustics. Proc VII Int Cong Exp Mech 1992:797–801
7. Konsta-Gdoutos M, Gdoutos EE (1992) Some remarks on caustics in mode-I stress intensity factor evaluation. Theor Appl Fract Mech 17:47–60
8. Konsta-Gdoutos M, Gdoutos EE (1992) Guidelines for applying the method of caustics in crack problems. Exp Tech 16:25–28
9. Konsta-Gdoutos M, Gdoutos EE (1992) Limit of applicability of the method of caustics in crack problems. Eng Fract Mech 42:251–263
10. Rosakis AJ (1993) Two optical techniques sensitive to gradients of optical path difference: the method of caustics and the coherent gradient sensor (CGS) In: Epstein JS (ed) VCH Publishers, pp 327–425
11. Etheridge JM, Dally JW (1978) A three-parameter method for determining stress intensity factors from isochromatic fringe patterns J Strain Anal 13:91–94
12. Etheridge JM, Dally JW, Kobayashi T (1978) A new method of determining stress intensity factor K from isochromatic fringe loops. Eng Fract Mech 10:81–93
13. Creager M, Paris P (1967) Elastic field equation for blunt cracks with reference to stress corrosion cracking. Int J Fract Mech 3:247–252
14. Theocaris PS (1970) Local yielding around a crack tip in plexiglas. J Appl Mech 92:409–415
15. Theocaris PS, Gdoutos EE (1972) An optical method for determining opening-mode and edge sliding-mode stress intensity factors. J Appl Mech 94:91–97
16. Theocaris PS (1972) Interaction between collinear asymmetric cracks. J Strain Anal 7:186–193
17. Theocaris PS (1972) Stress intensity factors at bifurcated cracks. J Mech Phys Sol 20:265–279
18. Theocaris PS, Gdoutos EE (1974) Verification of the validity of the Dugdale-Barenblatt model by the method of caustics. Eng Fract Mech 6:523–535
19. Theocaris PS, Gdoutos EE (1974) The modified Dugdale-Barenblatt model adapted to various fracture configurations in metals. Int J Fract 10:549–564
20. Theocaris PS (1976) Partly unbounded interfaces between dissimilar materials under normal and shear loading. Acta Mech 24:99–115
21. Gdoutos EE, Aifantis EC (1986) The method of caustics in environmental cracking. Eng Fract Mech 23:423–430
22. Katsamanis F, Raftopoulos D, Theocaris PS (1977) Static and dynamic stress intensity factors by the method of transmitted caustics. J Eng Mat 99:105–109
23. Kalthoff JF, Winkler S, Beinert J (1976) Dynamic stress intensity factors for arresting cracks in DCB specimens. Int J Fract 12:317–319
24. Theocaris PS, Gdoutos EE (1976) Surface topography by caustics. Appl Optics 15:1629–1638
25. Theocaris PS, Gdoutos EE (1977) Experimental solution of flexed plates by the method of caustics. J Appl Mech 44:107–111
26. Theocaris PS (1991) Elastic stress intensity factors evaluated by caustics. In: Sih GC (ed) Mechanics of fracture vol 7. Experimental evaluation of stress concentration and intensity factors, Martinus Nijhoff, pp 189–252
27. Kalthoff JF (1987) Shadow optical method of caustics. In: Kobayashi AS (ed) Handbook on experimental mechanics. Prentice Hall, Englewood Cliffs, pp 430–500

28. Gdoutos EE (2005) Fracture mechanics—an introduction, 3rd edn. Springer, Berlin
29. Theocaris PS, Papadopoulos GA (1981) Stress intensity factors from reflected caustics in birefringent plates with cracks. J Strain Anal 16:29–36
30. Dally JW (1979) Dynamic photoelastic studies of fracture. Exp Mech 19:349–361
31. Sanford RJ, Dally JW (1979) A general method for determining mixed mode stress intensity factors from isochromatic fringe patterns. J Eng Fract Mech 11:621–633
32. Creager M, Paris P (1967) Elastic field equation for blunt cracks with reference to stress corrosion cracking. Int J Fract Mech 3:247–252
33. Papadopoulos GA (1993) Fracture mechanics. The experimental method of caustics and the Det-criterion of fracture. Springer, Berlin
34. Lagarde A (1987) (ed) Static and dynamic photoelasticity and caustics. Recent developments. Springer, Berlin
35. Sridhar Krishnaswamy S (2000) Techniques for non-birefringent objects: Coherent shearing interferometry and caustics. In: Rastogi PK (ed) Photomechanics Topics Appl Phys 77. Springer, pp 295–321
36. Gdoutos EE (2016) The optical method of caustics. Opt Lasers Engng 79:68–77

Chapter 7
Photoelasticity

7.1 Introduction

Photoelasticity is a simple full-field optical method of experimental mechanics. Its name comes from the Greek words "photo" which means light and "elasticity" which refers to the ability of an object or material to resume its normal shape after being stretched or compressed. It is based on the phenomenon of **temporary or artificial double refractionor birefringence effect**, first discovered by Sir David Brewster in 1816, according to which transparent materials that are optically isotropic when unstressed become optically anisotropic and behave like birefringent crystals when they are stressed. The birefringence of the material is proportional to the difference of the principal stresses and is retained only during the application of loads. It disappears when the loads are removed.

The optical instrument of photoelasticity is the plane or circular polariscope that consists of a pair of crossed polarizers or a pair of crossed polarizers and two quarter-wave plates placed between the polarizers, respectively. The fringe patterns obtained in photoelasticity give the difference and the directions of the principal stresses. No special vibration-free optical bench is needed, as in other optical methods of experimental mechanics. A model of the structure is made and is tested in the laboratory. The method in the form of photoelastic coatings can be used for in-situ applications.

Photoelasticity uses non-coherent light. This is due to the fact that the optical path difference of the interfering light rays passing through the specimen is not much. Thus, interference between them easily takes place.

Photoelasticity has been successfully used for the analysis of stresses in engineering problems since the turn of the twentieth century. With the advent of the finite and boundary element method photoelasticity lost ground in engineering applications. However, it always retains its beauty, elegance, simplicity, and powerfulness.

In the following, we briefly present the fundamentals of photoelastic stress analysis including: optical patterns obtained in the plane and circular polariscopes, the isochromatic and isoclinic fringe patterns and their properties, the compensation methods, the stress separations methods, the fringe multiplication and sharpening methods, the transition from model to prototype, the three-dimensional photoelasticity and the photoelastic coatings. For more information, the interested reader is referred to the excellent books cited in the "references" at the end of this chapter.

7.2 Plane Polariscope

A **plane polariscope** consists of a pair of plane polarizers that have their optical axes parallel (light field) or perpendicular (dark field) to each other and a monochromatic or white light source (Fig. 7.1). Its name comes from using plane polarized light. The region between the two polarizers is referred to as the *field* of the polariscope. A model of the structure under study is inserted in the field of the polariscope. **Polarizer** is an optical element that divides an incident light beam into two mutually perpendicular components and allows the component parallel to its axis to be transmitted, while the component perpendicular to its axis to be absorbed. The polarizer close to the light source is called **polarizer**, while the second polarizer of the polariscope downstream the model is called **analyzer**. Usually, the dark-field arrangement (crossed polarizer and analyzer) of the plane polariscope is used.

We will analyze the optical transformations which take place in a dark field (no light passes through the analyzer) plane polariscope using the vectorial representation of polarized light and the Jones calculus (Sect. 2.8).

Consider the dark-field plane polariscope of Fig. 7.1 in which the axis of the polarizer is along the y-axis and the axis of the analyzer is along the x-axis. A linearly polarized light wave emerges from the polarizer with its light vector along the y-direction. It is given by Eq. (2.25)

$$E = a \cos \omega t \tag{7.1}$$

The light after leaving the polarizer enters the stressed model. According to the photoelastic law, it is resolved into two linearly polarized components E_1 and E_2 along the principal stress axes at the point of incidence. If β is the angle of one of the principal stresses with the y-axis E_1 and E_2 are given by

$$E_1 = E \cos \beta = a \cos \omega t \cos \beta$$
$$E_2 = E \sin \beta = a \cos \omega t \sin \beta \tag{7.2}$$

The components E_1 and E_2 propagate through the stressed model with different velocities. If δ_1 and δ_2 are the phase shifts of E_1 and E_2 with respect to a wave in air ($n = 1$), the two light waves after emerging from the model can be expressed as

$$E_1' = a \cos \beta \cos(\omega t - \delta_1)$$
$$E_2' = a \sin \beta \cos(\omega t - \delta_2) \tag{7.3}$$

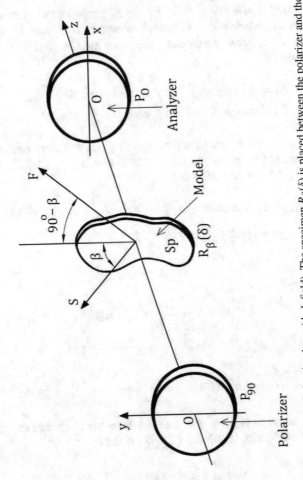

Fig. 7.1 Plane polariscope with crossed polarizer/analyzer (dark field). The specimen $R_\beta(\delta)$ is placed between the polarizer and the analyzer

where

$$\delta_1 = \frac{2\pi d}{\lambda}(n_1 - 1), \quad \delta_2 = \frac{2\pi d}{\lambda}(n_2 - 1) \tag{7.4}$$

n_1 and n_2 are the indices of refraction along the principal stress directions of the stressed model [given by Eq. (6.29)], d is the thickness of the model and λ is the wavelength of light.

After leaving the model, the two waves E_1' and E_2' propagate without any change and enter the analyzer, which allows only their components E_1'' and E_2'' along its x-axis to pass through. The components along the y-axis are absorbed by the analyzer. E_1'' and E_2'' are given by

$$\begin{aligned} E_1'' &= E_1' \sin \beta = a \cos \beta \cos(\omega t - \delta_1) \sin \beta \\ E_2'' &= E_2' \cos \beta = a \sin \beta \cos(\omega t - \delta_2) \cos \beta \end{aligned} \tag{7.5}$$

The components E_1'' and E_2'' are along the same x-plane and, therefore, they interfere. This is the case of a two-beam interference presented in Sect. 2.9.2. The resulting wave E_R that emerges from the analyzer is

$$\begin{aligned} E_R &= E_2'' - E_1'' = a \sin \beta \cos \beta [\cos(\omega t - \delta_2) - \cos(\omega t - \delta_1)] \\ &= a \sin 2\beta \sin \frac{\delta}{2} \sin\left(\omega t - \frac{\delta_2 + \delta_1}{2}\right) \end{aligned} \tag{7.6}$$

where

$$\delta = \delta_2 - \delta_1 = \frac{2\pi d}{\lambda}(n_2 - n_1) \tag{7.7}$$

Using Eq. (6.31) we obtain

$$\delta = \delta_2 - \delta_1 = \frac{2\pi d}{\lambda}(A - B)(\sigma_2 - \sigma_1) = \frac{2\pi d}{\lambda}c(\sigma_2 - \sigma_1), \quad c = A - B \tag{7.8}$$

The intensity I of the light wave that emerges from the analyzer is proportional to the square of its amplitude. Using Eq. (7.6) we obtain

$$I = k \sin^2 2\beta \sin^2 \frac{\delta}{2} \tag{7.9}$$

where k is a proportionality factor.

Equation (7.9) indicates that *the intensity of light that emerges from the analyzer in the plane polariscope depends on the difference of the principal stresses* $(\sigma_1 - \sigma_2)$ *and their angle of inclination* β.

Equation (7.9) was derived based on the vectorial representation of polarized light. We will now use the Jones calculus for the description of the optical transformations that take place in the plane polariscope. The Jones vector a of the linearly polarized light along the vertical axis of the polarizer is

$$a = \begin{bmatrix} 0 \\ 1 \end{bmatrix} \tag{7.10}$$

The Jones matrix $R_\beta(\delta)$ of the model, which is equivalent to a birefringent plate of birefringence δ and optical axis at an angle β with the x-axis, is

$$R_\beta(\delta) = \begin{bmatrix} e^{i\delta} \cos^2 \beta + \sin^2 \beta & (e^{i\delta} - 1) \sin \beta \cos \beta \\ (e^{i\delta} - 1) \sin \beta \cos \beta & e^{i\delta} \sin^2 \beta + \cos^2 \beta \end{bmatrix} \tag{7.11}$$

The Jones matrix of the analyzer with optical axis along the x-axis is

$$P = \begin{bmatrix} 1 & 0 \\ 0 & 0 \end{bmatrix} \tag{7.12}$$

According to the Jones calculus, the Jones vector a' of the light wave that emerges from the analyzer is

$$a' = PR_\beta(\delta)a = \begin{bmatrix} 1 & 0 \\ 0 & 0 \end{bmatrix} \begin{bmatrix} e^{i\delta} \cos^2 \beta + \sin^2 \beta & (e^{i\delta} - 1) \sin \beta \cos \beta \\ (e^{i\delta} - 1) \sin \beta \cos \beta & e^{i\delta} \sin^2 \beta + \cos^2 \beta \end{bmatrix} \begin{bmatrix} 0 \\ 1 \end{bmatrix}$$

$$= \begin{bmatrix} (e^{i\delta} - 1) \sin \beta \cos \beta \\ 0 \end{bmatrix} \tag{7.13}$$

The intensity I of the light wave a' is

$$I = k\tilde{a}'a \big[(e^{i\delta} - 1) \sin \beta \cos \beta \quad 0 \big] \begin{bmatrix} (e^{i\delta} - 1) \sin \beta \cos \beta \\ 0 \end{bmatrix}$$

$$= k\big[2 - (e^{i\delta} + e^{-i\delta}) \big] \sin^2 \beta \cos^2 \beta = k \sin^2 2\beta \sin^2 \frac{\delta}{2} \tag{7.14}$$

which is Eq. (7.9) derived using the vectorial representation of polarized light.

Equation (7.14) indicates that the intensity of light that emerges from the analyzer of a dark-field plane polariscope (the axes of polarizer and analyzer are perpendicular) becomes zero ($I = 0$) when

$$\beta = 0, \frac{\pi}{2}, \ldots, n\frac{\pi}{2} \quad \text{or} \quad \delta = 0, 2\pi, \ldots, n2\pi, \quad (n = 0, 1, 2, \ldots) \tag{7.15}$$

The first condition of light extinction ($\beta = 0, \frac{\pi}{2}, \ldots, n\frac{\pi}{2}$) is related to the directions of the principal stresses. The second condition of light extinction ($\delta = 0, 2\pi, \ldots, n2\pi$) is related to the difference of the stress-optical retardation of the model along the principal stress directions that is directly related to the difference of the principal stresses. We will discuss these two conditions separately in the sections below "Isoclinics" and "Isochromatics", respectively.

7.3 Circular Polariscope

A circular polariscope, as its name denotes, uses circularly polarized light produced by a combination of a polarizer and a quarter-wave plate with its optical axis at an angle 45° with the axis of the polarizer. A **quarter-wave plate** is an optical element that divides light into two linearly polarized beams along its two perpendicular axes with an optical path length difference between them equal to a quarter wave ($\lambda/4$). A quarter-wave plate corresponds to a particular wave length. A circular polariscope consists of the following four optical elements placed in sequence as following (Fig. 7.2):

i.   A linear polarizer, P_{90}, whose optical axis is vertical (along the y-axis).
ii.   A quarter-wave plate, Q_{45}, whose fast axis makes an angle 45° with the Ox-axis.
iii. A quarter-wave plate $Q_{\mp 45}$ whose fast axis makes an angle $-45°$ with the Ox-axis (for the dark-field polariscope the axes of the two quarter-wave plates are crossed) and $+45°$ with the Ox-axis (for the light-field circular polariscope the axes of the two quarter-wave plates are parallel).
iv.   A linear polarizer, called analyzer, P_0 whose optical axis is horizontal (along the x-axis).

Insertion of two quarter-wave plates into a plane polariscope converts it into a circular polariscope. The model is placed in the optical field of the circular polariscope between the two quarter-wave plates and is illuminated by circularly polarized

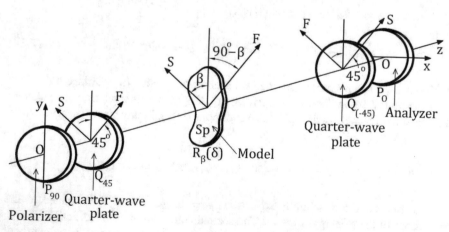

Fig. 7.2 Circular polariscope with dark field (crossed polarizer/analyzer and crossed quarter-wave plates). The specimen $R_\beta(\delta)$ is placed between the quarter-wave plates

light that exits the first quarter-wave plate after the polarizer. Two crossed quarter-wave plates (the fast axis of one is at 90° with the fast axis of the other) have no effect on the incident light. The light that exits the two quarter-wave plates is linearly polarized along the vertical axis of the polarizer. However, when the optical axes of the two quarter-wave plates are parallel (the fast axis of one is parallel to the fast axis of the other) the linearly polarized light that exits the second quarter-wave plates is linearly polarized along the horizontal axis of the analyzer. A circular polariscope with crossed quarter-wave plates and crossed polarizer and analyzer provides a dark field, while a circular polariscope with parallel quarter-wave plates and crossed polarizer and analyzer provides a light field. A stressed model is optically equivalent to a birefringent plate $R_\beta(\delta)$, with retardation δ and optical axis at angle β (the angle of one of the principal stresses) with the Ox-axis.

To obtain the optical effect downstream of the analyzer it is appropriate to use the Jones calculus that provides a simpler and more straightforward way of performing the optical transformations that take place in the circular polariscope than the vectorial representation of polarized light. We analyze the case of a dark-field circular polariscope with crossed quarter-wave plates and crossed polarizer/analyzer.

The Jones matrices Q_{45} and Q_{-45} of the quarter-wave plates with their optical axes at angles 45° and −45° with the x-axis are

$$Q_{45} = \frac{i+1}{2}\begin{bmatrix} 1 & i \\ i & 1 \end{bmatrix}, \quad Q_{-45} = \frac{i+1}{2}\begin{bmatrix} 1 & -i \\ -i & 1 \end{bmatrix} \tag{7.16}$$

The Jones vector a' of the light wave that emerges from the analyzer is (Fig. 7.2)

$$a' = P_0 Q_{-45} R_\beta(\delta) Q_{45} a$$

or

$$a' = \begin{bmatrix} 1 & 0 \\ 0 & 0 \end{bmatrix} \frac{i+1}{2}\begin{bmatrix} 1 & -i \\ -i & 1 \end{bmatrix} \begin{bmatrix} e^{i\delta}\cos^2\beta + \sin^2\beta & (e^{i\delta}-1)\sin\beta\cos\beta \\ (e^{i\delta}-1)\sin\beta\cos\beta & e^{i\delta}\sin^2\beta + \cos^2\beta \end{bmatrix}$$

$$\times \frac{i+1}{2}\begin{bmatrix} 1 & i \\ i & 1 \end{bmatrix}\begin{bmatrix} 0 \\ 1 \end{bmatrix} = \frac{1}{2}\begin{bmatrix} e^{i\delta}-1 \\ 0 \end{bmatrix} \tag{7.17}$$

The intensity of light with Jones vector a' is

$$I = k\bar{a}'a' = \frac{k}{4}[e^{i\delta}-1 \quad 0]\begin{bmatrix} e^{i\delta}-1 \\ 0 \end{bmatrix} = k\sin^2\frac{\delta}{2} \tag{7.18}$$

Equation (7.18) gives the intensity of light that exits the analyzer in a dark-field circular polariscope. Note that this equation is the same as that obtained for the dark-field plane polariscope with the exception that the term related to the angle of the principal stresses does not appear. Thus, the optical effect in the circular polariscope is independent of the angle of inclination of the principal stresses and depends only on the stress-optical retardation of the stressed model which is directly related to the difference of the principal stresses.

Consider now the case of a light-field circular polariscope with crossed polarizer/analyzer and quarter-wave plates with parallel axes. We have for the Jones vector a' of the light that exits the analyzer

$$a' = P_0 Q_{45} R_\beta(\delta) Q_{45} a \qquad (7.19)$$

or

$$a' = \begin{bmatrix} 1 & 0 \\ 0 & 0 \end{bmatrix} \frac{i+1}{2} \begin{bmatrix} 1 & i \\ i & 1 \end{bmatrix} \begin{bmatrix} e^{i\delta}\cos^2\beta + \sin^2\beta & (e^{i\delta}-1)\sin\beta\cos\beta \\ (e^{i\delta}-1)\sin\beta\cos\beta & e^{i\delta}\sin^2\beta + \cos^2\beta \end{bmatrix}$$
$$\times \frac{i+1}{2} \begin{bmatrix} 1 & i \\ i & 1 \end{bmatrix} \begin{bmatrix} 0 \\ 1 \end{bmatrix} = \frac{1}{2}\begin{bmatrix} e^{i\delta}+1 \\ 0 \end{bmatrix} \qquad (7.20)$$

The intensity of light with Jones vector a' is

$$I = k a' a' = \frac{k}{4}\begin{bmatrix} e^{-i\delta}+1 & 0 \end{bmatrix}\begin{bmatrix} e^{i\delta}+1 \\ 0 \end{bmatrix} = k\cos^2\frac{\delta}{2} \qquad (7.21)$$

Equation (7.21) gives the intensity of light that exits the analyzer in a light-field circular polariscope. It is similar to Eq. (7.18) with the sinus term replaced by a cosine term. It indicates that the optical effect downstream of the analyzer depends only on the stress-optical retardation of the stressed model which is linearly related to the difference of the principal stresses.

7.4 Isoclinics

The condition of extinction of light ($I = 0$) downstream the analyzer of a dark-field plane polariscope when $\beta = 0, \pi/2, \ldots, n\,\pi/2$ developed previously indicates that the points of the model at which the directions of the principal stresses coincide with the directions of the crossed polarizer/analyzer appear dark. This result should be expected, since, when the axis of the polarizer coincides with the direction of one of the principal stresses the linearly polarized wave exits the model at the direction of the polarizer. This wave meets the axis of the analyzer at normal angle and, therefore, no light is transmitted through the analyzer. *The loci of points at which the principal stress directions coincide with the crossed directions of the polarizer/analyzer belong to the same dark fringe.* These fringes are called **isoclinics**. The name comes from the Greek words, "iso" which means the "same" and "clinic", which means inclination, that is, **lines of same inclination** of the principal stresses.

From the analysis of the optical patterns obtained in the plane and circular polariscopes, we observe that the isoclinics appear only in the plane polariscope, and disappear in the circular polariscope.

7.5 Isochromatics

Extinction of light ($I = 0$) downstream the analyzer of a dark-field plane polariscope or a dark-field circular polariscope when $\delta = 0, 2\pi, ..., n\,2\pi\ (n = 0, 1, 2, ...)$ developed previously indicates that the points of the body which satisfy this condition appear dark. These points belong to the same fringe. Therefore, *the loci of points at which the difference of the principal stresses is zero or produces an integral number of wavelengths of retardation* ($\delta = 0, 2\pi, ..., n\,2\pi$)*belong to the same dark fringe.* These fringes are called **isochromatics** of integer order. The name comes from the Greek words, "iso" which means the "same" and "chromatics" which means color, that is, **lines of same color**. Isochromatics appear in the same color when white light is used.

From the previous analysis of the light intensity in a bright-field circular (or plane) polariscope it is concluded that the points of the body satisfy the condition $\delta = (1+2n)\pi, (n = 0, 1, 2, ...)$ appear dark. These points belong to the same fringe. Therefore, *the loci of points at which the difference of the principal stresses produces an even number of half-wavelengths of retardation* ($\delta = \pi, 3\pi, ... (1 + 2n)\pi, n = 0, 1, 2, ...$)*belong to the same fringe.* These fringes are the isochromatics of half order.

From the above analysis, we conclude that *in a plane polariscope we obtain two superimposed families of curves, the isoclinics, and the isochromatics.* The isoclinics provide the directions of the principal stresses and the isochromatics provide the difference of the principal stresses. In a circular polariscope, the isoclinics disappear. We obtain only integer-order isochromatics with stress-optical retardation $\delta = 0, 2, \pi, ..., n\,2\pi\ (n = 0, 1, 2, ...)$ in the dark-filed circular polariscope and half-order isochromatics with stress-optical retardation $\delta = \pi, 3\pi, ... (1+2n)\pi, (n = 0, 1, 2, ...)$ in the light-field circular polariscope. Both integer-order and half-order isochromatics appear in the dark-field and bright-field plane polariscope, respectively.

From Eq. (7.21) we obtain for the integer fringe order N

$$N = \frac{\delta}{2\pi} = n \tag{7.22}$$

where N refers to the number of full phase or wavelength cycles. For example, if the phase difference $\varphi = 2\pi$ or the wavelength difference is λ the fringe order is one (*1*), if $\varphi = 4\pi$ or the wavelength difference is 2λ the fringe order is two (2), etc. Using Eq. (7.8) we obtain

$$n = N = \frac{\delta}{2\pi} = \frac{d}{\lambda}c(\sigma_l - \sigma_2) \tag{7.23}$$

Equation (7.23) indicates that the fringe order N depends on the difference of the principal stresses $(\sigma_1 - \sigma_2)$, the stress-optical constant of the material c, the thickness of the model d, and the wavelength λ of the monochromatic light used in the experiment. For a given wavelength λ we introduce the **photoelastic constant** f_σ by

$$f_\sigma = \frac{\lambda}{c}$$

(7.24)

and Eq. (7.23) becomes

$$n = N = \frac{d}{f_\sigma}(\sigma_1 - \sigma_2),$$

(7.25)

or

$$\sigma_1 - \sigma_2 = \frac{N f_\sigma}{d}$$

(7.26)

Equation (7.24) indicates that the photoelastic constant f_σ depends on the stress-optical constant of the material c and the wavelength λ of the light. Thus, f_σ is a *material constant at a given wavelength*. f_σ is measured from a calibration test. Equation (7.26) is the *basic equation of two-dimensional photoelasticity*. It provides the difference of the principal stresses at a point when the fringe order N is known at that point.

Equation (7.23) gives the integer-order isochromatics for the dark-field circular polariscope. For the light-field circular polariscope we have for the stress-optical path retardation

$$\delta = (1 + 2n)\pi, \quad (n = 0, 1, 2, \ldots)$$

and

$$N = \frac{\delta}{2\pi} = \frac{1}{2} + n, \quad (n = 0, 1, 2, \ldots)$$

(7.27)

Equation (7.27) indicates that in light-field circular polariscope isochromatics of orders 1/2, 3/2... $n + 1/2$ $(n = 0, 1, 2, \ldots)$ are obtained. *Thus, the dark- or light-field circular polariscope provides integer-order 1, 2, ... n or half-order 1/2, 3/2 ... n + 1/2 isochromatics, respectively only. Integer-order and half-order isochromatics appear in the dark-field and bright-field plane polariscope, respectively, superposed with the isoclinics.*

7.6 Isochromatics with White Light

The isochromatic fringes in the previous analysis of a plane or circular polariscope obtained with monochromatic light appear dark. Consider now that white light is used. It consists of all wavelengths of the visible spectrum (400–800 nm). In Eq. (7.23) the material constant c is independent of the wavelength. For a particular wavelength λ (which corresponds to a given color) there is a value of $(\sigma_1 - \sigma_2)$ which satisfies Eq. (7.23). When a color is extinguished from the white color spectrum its complementary color appears. The isochromatic fringe at a point the light intensity becomes zero does not appear dark but colored. It has the complementary color of the color that is extinguished. Thus, *the isochromatic fringe pattern for white light consists of colored fringes. The colors in the fringe pattern are the complementary colors of the colors extinguished.*

Approximate values of the wavelengths of colors of the visible spectrum with their complementary colors are listed in Table 7.1. The complimentary colors of a first extinction appear in the order: black, white, pale yellow, orange, deep red, purple, deep blue, blue green. By increasing $(\sigma_1 - \sigma_2)$ beyond this cycle the colors begin to reappear in the same order. For a second green, the values of $(\sigma_1 - \sigma_2)$ is twice that of the first green, for the third green three times that of the first, etc. The first green marks the end of the first cycle. The colors between the initial black and the green are the colors of the first cycle.

As $(\sigma_1 - \sigma_2)$ increases after the first cycle, the colors are extinguished a second time in the same order as in the first cycle. However, the complementary colors obtained during the second cycle are not exactly the same as in the first cycle. This is because during the second cycle additional colors are extinguished, which were

Table. 7.1 Approximate values of the wavelengths of colors of the visible spectrum with their complementary colors

Color extinguished	Complementary color appearing in pattern	Approximate retardation (Å)
All colors	Black	0
None	White	2600
Extreme Violet*	Pale Yellow	3500
Blue*	Orange	4600
Green*	Deep Red	5200
Yellow*	Purple	5900
Orange*	Deep Blue	6200
Red*	Blue Green	7000
Red and Violet**	Yellow Green	8000
Bleu**	Orange	9400
Green**	Rose Red	10,500
Yellow**	Purple	11,500
Red and Violet**	Green	13,000

not extinguished during the first cycle and the extinctions of the new colors coincide approximately with the second or third extinctions of former colors. For example, the wavelength of violet is 390 nm and that of deep red is 770 nm, that is λ of deep red is approximately equal to twice the λ of violet. Thus, the second extinction of violet coincides approximately with the first extinction of deep red. Also, the second extinction of ordinary red with $\lambda = 650$ nm coincides with the third extinction of indigo with $\lambda = 433$ nm. Thus, the colors in the second cycle are somewhat modified by the colors of the first cycle, and so on. The nearly simultaneous extinction of the second violet with the first deep red gives a deep pink, and the simultaneous extinction of the second red and the third indigo gives a predominant green. The second green which is followed by approximately white light ends the second cycle. In the higher cycles yellowish white, pink, and green follow one another. These colors get progressively paler because of the overlapping of the complementary colors.

White light can be used to increase the accuracy of the determination of $(\sigma_1 - \sigma_2)$ in regions of low birefringence by a color-matching technique. Use of white light is useful in identifying the zero-order isochromatic since they appear black.

7.7 Properties of Isoclinics

Isoclinics are the loci of points the principal stresses have the same direction which coincides with the crossed direction of the axes of polarizer/analyzer. They are obtained only in the plane polariscope by rotating the crossed pair of polarizer/analyzer. The angle the axis of the polarizer or analyzer makes with the x- or y-axis is the **parameter of the isoclinic**. They are usually obtained at angle intervals of 10^0. In the circular polariscope, the isoclinics disappear. Isoclinics are black, with either monochromatic or white light. It is recommended to obtain isoclinics with white light instead of monochromatic light to increase their distinctness.

Isoclinics are distinguished from isochromatics as they appear black in white light, while isochromatics appear colored. They are steady when the load is changed, while isochromatics move. Usually, isoclinics are obtained using optically inert materials (with low birefringence), like Plexiglas, at low loads to diminish the appearance of isochromatics.

Isoclinics are used to obtain the stress trajectories (orthogonal families of curves across which the principal stresses are tangent) (Fig. 7.3). A fundamental theorem of stress trajectories is: *if a system of stress trajectories be divided into two orthogonal families of curves, then all the stresses tangent to one family are maximum stresses and those tangent to the other family are minimum stresses* [1].

Here are some properties of isoclinics useful in photoelastic and stress analysis:

i. Isoclinics do not intersect each other (there is only one direction of principal stresses at a point).

    i. The axis of symmetry of a symmetric body loaded symmetrically with regard to this axis belongs to an isoclinic of one parameter. The isoclinic

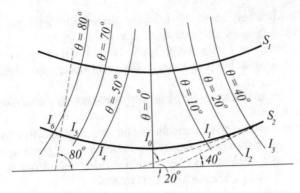

Fig. 7.3 Isoclinics $I_0, I_1, I_2, I_3, I_4, I_5, I_6$, of angles 0°, 10°, 20°, 40°, 50°, 70°, 80°, respectively, and stress trajectories S_1 and S_2

       containing the axes of symmetry may consist of other branches in addition to the line of symmetry.

ii.   The parameter of an isoclinic at the point of intersection of a free boundary is determined by the slope of the boundary at the point of intersection (at free boundaries there is only one principal stress tangent to the boundary, the principal stress perpendicular to the boundary is zero).

iii.   Isoclinics pass through points at which the principal stresses are equal (at these points all directions are principal stress directions).

iv.   Isoclinics pass through points of concentrated loads (at concentrated loads all directions are principal stress directions).

v.   The principal stresses tangent to a given stress trajectory are a maximum or minimum where an isoclinic cuts the stress trajectory at right angles. As a result, an isoclinic must be normal to a free boundary where the boundary stress is a maximum or minimum [1].

7.8 Properties of Isochromatics

Isochromatics are the loci of points of equal value of the difference of principal stresses $(\sigma_1 - \sigma_2)$ or of the maximum in-plane shear stress ($\tau_{max} = (\sigma_1 - \sigma_2)/2$). They are obtained by the plane or the circular polariscope. In the plane polariscope the isochromatics are obtained together with the isoclinics, while in the circular polariscope only the isochromatics are obtained (the isoclinics disappear).

    Here are some properties of isochromatics useful in photoelastic and stress analysis:

ii.   Isochromatics do not intersect each other (there is only one value of principal stresses at a point).

iii.   *The boundary stresses (parallel to the boundary) are directly determined from the isochromatic fringe pattern* (the stress normal to the free boundary is zero). This is a very important result of photoelastic analysis since, usually, the boundary stresses are the highest stresses in a body.

iv. The fringe order at a point for an applied load is determined by counting the number of fringes that pass through the point during the application of the load.

v. Black dots in the isochromatic pattern appear at points where the two principal stresses are equal.

vi. In-plane polariscope with monochromatic or white light isoclinics and zero-order isochromatics appear black. Isochromatics at a given applied load are separated from isoclinics as they remain steady, while isoclinic move as the system of crossed polarizer/analyzer is rotated.

vii. In-plane polariscope the isochromatics are distinguished from the isoclinics when white light is used. The isochromatics appear colored, while the isoclinics appear black.

7.9 Compensation Methods

7.9.1 Introduction

Isochromatic fringes of ½ wavelength (λ) can be determined by a dark- and light-field circular polariscope. For higher accuracy, we resort to the so-called **compensation methods**. In these methods, the directions of the principal stresses must be known beforehand. The methods are point-by-point. They can determine the fringe order with an accuracy of the order of 1% wavelength.

In the following, we briefly present the tension/compression specimen used as a compensator, the Babinet and the Babinet-Soleil compensators which use optical elements, and the Senarmont and Tardy compensation methods. The first group of compensators uses an external element to zero the stress-optical retardation at the point under consideration, while the Senarmont and Tardy compensation methods use modifications of the circular polariscope.

7.9.2 The Tension/Compression Specimen

The principle of using a tension/compression specimen as a compensator relies on the fact that a plane state of stress can be considered as a superposition of a hydrostatic and a uniaxial state of stress. A hydrostatic state of stress does not produce an optical effect in the circular polariscope. A uniaxial stress state, tension or compression, can either be nullified or transformed to a hydrostatic state by superimposing a tension/compression specimen along the direction of the uniaxial stress or along the perpendicular direction. In both cases, the optical effect is zero. By knowing the applied tension/compression of an external transparent specimen the retardation at the point under consideration is determined.

For the application of the tension/compression specimen as a compensator, the directions of the principal stresses at the point of interest are determined from the order of isoclinic that passes from that point. The tension or compression specimen is placed along or perpendicular to the direction of one of the principal stresses and the load is increased until the optical effect is zero. Then, from the value of the stress of the tension or compression specimen, the difference of the principal stresses at the point is determined.

7.9.3 Babinet and Babinet-Soleil Compensators

Both compensators are based on the same principle as the tension/compression specimen. However, in this case, the necessary external retardation for the nullification of the optical effect is provided through an optical device. In the Babinet method, the optical device consists of two acute quartz prisms or wedges with perpendicular optic axes which are adjusted to provide the retardation needed to nullify the existing retardation at the point considered. In the Babinet-Soleil compensator, the external retardation is provided by two wedges of additive effect and an additional quartz plate with its optic axis perpendicular to the optic axis of the wedges. Both compensators can measure retardations of 1% wavelength.

7.9.4 Sernarmont Compensation Method

The Sernarmont compensation method uses the optical elements of the circular polariscope arranged as follows: the first quarter-wave plate is removed, the axis of polarizer along the Oy-axis is placed at 45° with the principal stress directions at the point under consideration, one axis of the second quarter-wave plate is arranged parallel to the axis of the polarizer and the axis of the analyzer makes an angle θ with the Ox-axis. (Fig. 7.4).

The Jones vector a' of the light wave that emerges from the quarter-wave plate is given by

$$a' = Q_0 R_{4,5}(\delta)a \tag{7.28}$$

where:

Q_0 = Jones matrix of the quarter-wave plate

$R_{45}(\delta)$ = Jones matrix of the model

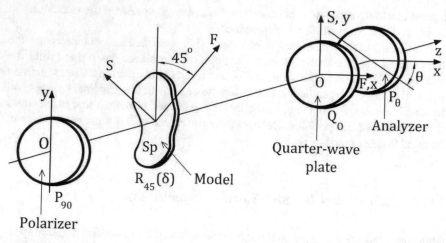

Fig. 7.4 Arrangement of optical elements of the circular polariscope for the Senarmont compensation method

a = Jones vector of the linearly polarized light that emerges from the polarizer with vertical plane of vibration

Equation (7.28) can be written as

$$a' = Q_0 R_{4,5}(\delta)a = \begin{bmatrix} i & 0 \\ 0 & 1 \end{bmatrix} \frac{1}{2} \begin{bmatrix} e^{i\delta} + 1 & e^{i\delta} - 1 \\ e^{i\delta} - 1 & e^{i\delta} + 1 \end{bmatrix} \begin{bmatrix} 0 \\ 1 \end{bmatrix}$$

$$= e^{i\delta/2} \begin{bmatrix} -\sin\left(\frac{\delta}{2}\right) \\ \cos\left(\frac{\delta}{2}\right) \end{bmatrix} = e^{i\delta/2} \begin{bmatrix} \cos\left(\frac{\pi+\delta}{2}\right) \\ \sin\left(\frac{\pi+\delta}{2}\right) \end{bmatrix} \tag{7.29}$$

Equation (7.28) indicates that the light wave that emerges from the quarter-wave plate downstream the model (the other quarter-wave plate was removed) is linearly polarized with plane of polarization at an angle $\delta/2$ with the vertical axis. Thus, if the analyzer is rotated counterclockwise through an angle $\theta = \delta/2$ its principal axis becomes perpendicular to the linearly polarized light that emerges from the quarter-wave plate, and extinction occurs. The relative retardation at the point considered is

$$N = k + \frac{\theta}{\pi}, \quad (k = 1, 2, \ldots, n) \tag{7.30}$$

Equation (7.30) *expresses the fractional fringe order according to the Senarmont compensation method.*

7.9.5 *Tardy Compensation Method*

In the Tardy compensation method, no element of the circular polariscope is removed. The system of polarizer, analyzer, and two quarter-wave plates is rotated until the axes of polarizer and analyzer become parallel to the principal stress directions at the point considered (Fig. 7.5).

The Jones vector a' of the light wave that emerges from the quarter-wave plate in front of the analyzer is given by

$$a' = Q_{-45}\, R_0(\delta)\, Q_{45} a \qquad (7.31)$$

where:

$Q_{\pm 45}$ = Jones matrix of the quarter-wave plate whose fast axis makes an angle $\pm 45°$ with the Ox-axis.

$R_0(\delta)$ = Jones matrix of the model whose principal stress direction is parallel to the x-axis

a = Jones vector of the linearly polarized light that emerges from the polarizer with vertical plane of vibration

Equation (7.31) is written as

$$a' = Q_{-45}\, R_Q(\delta)\, Q_{45} a = \left(\frac{1+i}{2}\right)^2 \begin{bmatrix} 1 & -i \\ -i & 1 \end{bmatrix}\begin{bmatrix} e^{i\delta} & 0 \\ 0 & 1 \end{bmatrix}\begin{bmatrix} 1 & i \\ i & 1 \end{bmatrix}\begin{bmatrix} 0 \\ 1 \end{bmatrix}$$

$$= \left(\frac{1+i}{2}\right)^2 e^{i\delta/2}\begin{bmatrix} \cos\left(\frac{\pi+\delta}{2}\right) \\ \sin\left(\frac{\pi+\delta}{2}\right) \end{bmatrix} \qquad (7.32)$$

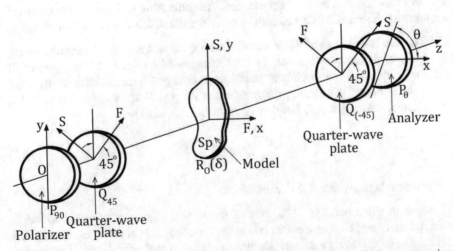

Fig. 7.5 Arrangement of optical elements of the circular polariscope for the Tardy compensation method

Equation (7.28) indicates that the light wave that emerges from the quarter-wave plate is linearly polarized with plane of polarization at an angle $\delta/2$ with the vertical *(Oy)* axis, as in the case of the Senarmont compensation method. Thus, if the analyzer is rotated counterclockwise through an angle $\theta = \delta/2$ its principal axis becomes perpendicular to the linearly polarized light that emerges from the quarter-wave plate, and extinction occurs. The relative retardation at the point considered is given by Eq. (7.30).

Equation (7.30)*expresses the fractional fringe order according to the Tardy compensation method*. Note that this equation is the same as that of the Senarmont compensation method. Accuracies of ∓ 0.02 fringes can be reached by both compensation methods.

7.10 Determination of the Photoelastic Constant f_s

Determination of the difference of the principal stresses $(\sigma_1 - \sigma_2)$ from the isochromatic fringe pattern using Eq. (7.26) requires the value of the photoelastic constant f_s. When the stress-optical constant c and the wavelength λ of the light used in the plane or circular polariscope are known f_s can be calculated from Eq. (7.24). Values of c for photoelastic materials are given in the literature. However, these values may vary with age, temperature, batch, etc. It is, therefore, necessary to determine the value of f_s at the time of the experiment using a calibration test.

In principle, any specimen for which the stress state at a point is known either theoretically or experimentally can serve as calibrating specimen. The specimen should be easy to make and test. It is loaded incrementally and the load and the fringe order are recorded simultaneously. From the values of load and fringe order, the photoelastic constant f_s is determined. As calibrating test, we will present the tension test, the beam in pure bending, and the circular disk in diametral compression.

i. **Tension test**: The state of stress in a tension specimen is uniform tension at every point of the specimen. When the load is increased incrementally the surface of the body in a plane or circular polariscope changes color simultaneously from light to dark. By counting the number of fringes N at an applied load P the photoelastic constant f_s is determined from Eq. (7.26) as

$$f_s = \frac{1}{b}\left(\frac{P}{N}\right) \tag{7.33}$$

where b is the width of the specimen.

Beam in Pure bending: The state of stress in a beam under pure bending is uniaxial tension or compression for points on opposite sides of the neutral axis of the beam. The isochromatics are straight lines parallel to the neutral axis of the beam. The zero fringe order coincides with the neutral axis (the mid fiber of the beam). From the fringe order N at distance y from the neutral axis, the photoelastic constant

f_s is determined from Eq. (7.26) as

$$f_s = \frac{12y}{h^3}\left(\frac{M}{N}\right) \tag{7.34}$$

where M is the applied bending moment and h is the height of the beam.

Circular disk in diametral compression. Closed form solution of the state of stress in a disk subjected to diametral compression is available in elasticity books. Measurements are usually made at the center of the disk. At this point the difference of principal stresses is

$$\sigma_1 - \sigma_2 = \frac{8P}{\pi dD} \tag{7.35}$$

where P is the load, D is the diameter and d is the thickness of the disk.

The photoelastic constant f_s is determined from Eq. (7.26) as

$$f_s = \frac{8}{\pi D}\left(\frac{P}{N}\right) \tag{7.36}$$

Note that in Eqs. (7.33), (7.34), and (7.36) the value of f_s is independent of the thickness of the specimen.

7.11 Stress Separation

Photoelastic data provide two stress related quantities:

i. **The difference of the principal stresses** obtained from the isochromatic fringe pattern using the plane or circular polariscope.
ii. **The directions of the principal stresses** obtained from the isoclinics using the plane polariscope.

For stress free boundaries the normal stress perpendicular to the boundary and the shear stress is zero. Thus, the normal stress parallel to the tangent of the boundary is determined from the isochromatic fringe pattern. Since maximum stresses usually appear along boundaries, **photoelasticity directly provides the maximum stresses**.

From photoelastic data obtained from isochromatics and isoclinics, the in-plane shear stress is calculated as

$$\tau_{xy} = -\frac{\sigma_1 - \sigma_2}{2}\sin 2\beta \tag{7.37}$$

where β is the angle between the direction of principal stresses σ_1 or σ_2 and the x-axis.

For the determination of the individual principal stresses σ_1, σ_2 a few methods have been developed. We will briefly present the shear-difference method and methods for the determination of the sum of the principal stresses.

The shear-difference method. The method uses Eq. (7.37) for the determination of the shear stress τ_{xy} from photoelastic data in combination with the equations of equilibrium. The equation of equilibrium along the x-axis (by omitting the body forces) is

$$\frac{\partial \sigma_x}{\partial x} + \frac{\partial \tau_{xy}}{\partial y} = 0 \tag{7.38}$$

Integration of this equation along the x-axis between two points O and i gives

$$\sigma_x^i = \sigma_x^o - \int_0^i \frac{\partial \tau_{xy}}{\partial y} dx \tag{7.39}$$

where σ_x^i and σ_x^o are the values of σ_x at points i and O of the x-axis.
    For points 1 and 2 we have

$$\frac{\partial \tau_{xy}}{\partial y} = \frac{\Delta \tau_{xy}}{\Delta y} = \frac{(\tau_{xy})_1 - (\tau_{xy})_2}{\Delta y} \tag{7.40}$$

where $(\tau_{xy})_1$ and $(\tau_{xy})_2$ are the values of τ_{xy} at points 1 and 2.
    For a small distance, $\Delta x = x_1 - x_2$ between points 1 and 2 Eq. (7.39) becomes

$$\sigma_x^2 = \sigma_x^1 - \Delta \tau_{xy} \left(\frac{\Delta x}{\Delta y} \right) \tag{7.41}$$

For $\Delta x = \Delta y$, we obtain

$$\sigma_x^2 = \sigma_x^1 - \Delta \tau_{xy} \tag{7.42}$$

Equation (7.41) or (7.42) is used for the determination of the normal stress σ_x^2 at point 2 when the normal stress σ_x^1 at point 1 and $\Delta \tau_{xy}$ are known.
    For the application of the method, the normal stress at point O is first determined (when the point belongs to a free boundary it is determined from the isochromatic pattern). Then, $\Delta \tau_{xy}$ is determined by calculating from the isocromatic and isoclinic patterns the shear stress τ_{xy} along two lines parallel to the x-axis and at distances $\Delta y/2$. Finally, we apply Eq. (7.41) or (7.42) incrementally along the x-axis.
    When the normal stress σ_x is known the other normal stress σ_y is determined by the following equation

$$\sigma_y = \sigma_x \pm \sqrt{(\sigma_1 - \sigma_2)^2 - 4\tau_{xy}^2} \tag{7.43}$$

Methods for determining the sum of the principal stresses. The individual principal stresses σ_1, σ_2 can be determined when their difference (calculated from the isochromatic fringe pattern) is combined with their sum. In the following, we will present numerical and experimental methods for the determination of the sum of the principal stresses.

i. **Numerical methods**. The sum of the principal stresses $(\sigma_1 + \sigma_2)$ satisfies the harmonic (Laplace) equation (when $\nabla^2 \Omega = 0$, Ω is the body force potential)

$$\nabla^2(\sigma_1 + \sigma_2) = \left(\frac{\partial^2}{\partial x^2} + \frac{\partial^2}{\partial y^2} \right)(\sigma_1 + \sigma_2) = 0 \qquad (7.44)$$

The values of $(\sigma_1 + \sigma_2)$ along a free boundary are determined from the isochromatic pattern (the principal stress normal to the boundary is zero, $\sigma_2 = 0$, so $\sigma_1 + \sigma_2 = \sigma_1 - \sigma_2 = \sigma_1$). Numerical methods have been developed for the solution of Eq. (7.44) with known boundary conditions (**Dirichlet problem**). A simple numerical scheme is based on finite differences.

A function φ that satisfies Eq. (7.44) can be put in finite form as

$$\varphi_0 = \frac{\varphi_1 + \varphi_2 + \varphi_3 + \varphi_4}{4} \qquad (7.45)$$

where φ_0 is the value of the function φ at a point O and φ_i $(i = 1, 2, 3, 4)$ are the values of the function at four equidistant neighboring points of an orthogonal grid.

Equation (7.45) can be applied to interior points of a mess in the region. By knowing the boundary values of φ its values at interior points can be determined from the solution of a system of linear equations.

ii .**Experimental methods**. In many cases of problems in physics, different systems are described by the same equations. Due to the analogy among the systems, the behavior of one can be determined from the behavior of the other, which may be easier to solve. We will present the elastic membrane, the electrical analogy, and methods using the change of the thickness of the model for the determination of the sum of the principal stresses $(\sigma_1 + \sigma_2)$.

The elastic membrane analogy method, also known as soap-film analogy method is based on the satisfaction of the Laplace equation by the deflections of an elastic thin membrane uniformly stretched along its boundary. We construct a box in the shape of the model under study with heights along its boundary proportional to $(\sigma_1 + \sigma_2)$ and stretch a membrane along its boundary. The height of the membrane at any point of the model is proportional to $(\sigma_1 + \sigma_2)$. The constant of proportionality is that used for the heights of the box along its boundary. To measure the heights of the membrane the out-of-plane shadow moiré method can be used. A grating is placed in front of the membrane and the surface of the membrane is made matt reflective. The interference of the grating with its shadow from the membrane creates fringes which are the loci of points of equal deflections of the membrane.

The **electrical analogy method** is based on the satisfaction of the Laplace equation by the potential of an electrostatic field. By applying voltages proportional to (σ_1 + σ_2) along the boundary of a conducting medium having the shape of the model, the resulting voltages at points inside the model are proportional to ($\sigma_1 + \sigma_2$). The voltage is applied along the boundary of the medium with contacts. The number and width of the contacts depend on the variation of ($\sigma_1 + \sigma_2$) along the boundary. More contacts are placed at positions of steep variation of stresses.

The methods using **the change of thickness Δd of the model** are based on the linear relationship between Δd and ($\sigma_1 + \sigma_2$) under conditions of plane stress. Δd is usually measured by interferometric methods, as we will discuss in the next chapter.

7.12 Fringe Multiplication and Sharpening

The number of isochromatics in a photoelastic experiment is dictated by the applied load, the thickness, and the photoelastic constant of the model material. There is a limitation in the values of the above three quantities. The applied load cannot exceed some value because the maximum stress in the model should be below the elastic limit. The photoelastic constant and the thickness of the model are dictated by the material and the requirement that experiments are performed under conditions of plane stress. There may exist regions in the model of sparse isochromatics. Furthermore, isochromatics and isoclinics may appear broad, and their exact center-line cannot be accurately determined. All these factors influence the accuracy of the interpretation of the photoelastic data. For the accurate determination of the isochromatic fringe order at a point various compensation methods can be used. However, these are point-by-point methods and do not provide full-field data.

Post [34, 35] developed a method for fringe multiplication and sharpening by inserting two partial slightly inclined mirrors on both sides of the specimen into a lens circular polariscope (Fig. 7.6). The mirrors allow the light to pass through the

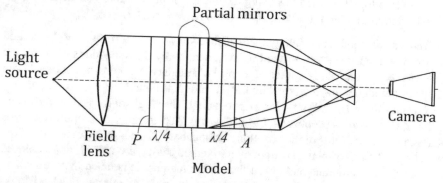

Fig. 7.6 Fringe multiplication by inserting two partial mirrors in the circular polariscope

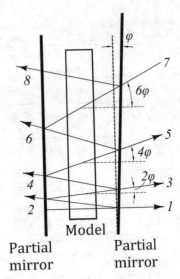

Fig. 7.7 Light reflection and transmission between two slightly inclined partial mirrors for fringe multiplication

specimen a number of times resulting in fringe multiplication (Fig. 7.7). Usually, multiplication of 5–7 times is achieved. After that, a considerable loss of light intensity occurs. By this optical arrangement, not only fringe multiplication but also fringe sharpening occurs. The light intensity distribution between two fringes is not sinusoidal, as is the case without the two mirrors, and presents a sharper distribution which allows the more accurate location of the centerlines of the fringes.

7.13 Transition From Model to Prototype

In photoelasticity, a model of the prototype is made at a given scale and the results of stress analysis are transferred from the model to the prototype. The elastic constants (modulus of elasticity E and Poisson's ratio v) of the model material differ from those of the prototype material. The question arises of how to transfer the stresses from the model to the prototype. The stresses in two-dimensional bodies under conditions of plane stress or plane strain are independent of E. Concerning the dependence of stresses on v the following rules apply:

i.   **For simply connected bodies** (without holes). The stresses are independent of Poisson's ratio v when the body forces are either absent or vary linearly in the body.

ii. **For multiple-connected bodies** (with holes). The stresses are independent of Poisson's ratio v when the resultant force acting on the boundary of the hole

Table 7.2 Scaling equations for transferring stresses, strains, and deflections from model (m) into prototype (p) for linear elastic bodies. The effect of Poisson's ratio is considered negligible

	To find prototype stresses	To find prototype strains	To find prototype deflections
If moments are given	$\sigma_p = \lambda^3 \frac{M_p}{M_m} \sigma_m$	$\varepsilon_p = \lambda^3 \frac{M_p}{M_m} \frac{E_m}{E_p} \varepsilon_m$	$u_p = \lambda^3 \frac{M_p}{M_m} \frac{E_m}{E_p} u_m$
If concentrated loads are given	$\sigma_p = \lambda^2 \frac{P_p}{P_m} \sigma_m$	$\varepsilon_p = \lambda^2 \frac{P_p}{P_m} \frac{E_m}{E_p} \varepsilon_m$	$u_p = \lambda \frac{P_p}{P_m} \frac{E_m}{E_p} u_m$
If loads per unit length are given	$\sigma_p = \lambda \frac{Q_p}{Q_m} \sigma_m$	$\varepsilon_p = \lambda \frac{Q_p}{Q_m} \frac{E_m}{E_p} \varepsilon_m$	$u_p = \frac{Q_p}{Q_m} \frac{E_m}{E_p} u_m$
If loads per unit surface are given	$\sigma_p = \frac{R_p}{R_m} \sigma_m$	$\varepsilon_p = \frac{R_p}{R_m} \frac{E_m}{E_p} \varepsilon_m$	$u_p = \frac{1}{\lambda} \frac{M_p}{M_m} \frac{E_m}{E_p} u_m$
If loads per unit volume are given	$\sigma_p = \frac{1}{\lambda} \frac{S_p}{S_m} \sigma_m$	$\varepsilon_p = \frac{1}{\lambda} \frac{S_p}{S_m} \frac{E_m}{E_p} \varepsilon_m$	$u_p = \frac{1}{\lambda^2} \frac{M_p}{M_m} \frac{E_m}{E_p} u_m$

vanishes or reduces to a couple. Otherwise, the stresses depend on v and a correction is needed.

Scaling laws for transferring stresses from model to prototype are based on the so-called Buckingham π theorem. If a model is manufactured geometrically similar to the prototype, stresses at points of the prototype can be obtained from stresses at corresponding points of the model from

$$\sigma_p = \sigma_m \frac{L_m}{L_p} \frac{d_m}{d_p} \frac{P_p}{P_m} \qquad (7.46)$$

where σ is the stress, L is a linear dimension, d is the thickness and P is the applied load. Quantities with subscripts p and m refer to the prototype and model, respectively.

Table 7.2 presents scaling equations for transferring stresses, strains, and deflections from model into prototype for linear elastic bodies. The effect of Poisson's ratio is considered negligible. λ is the ratio of a length in model and the corresponding length in prototype.

7.14 Three-Dimensional Photoelasticity

In three-dimensional bodies, the state of stress is defined by six stresses. Solution of three-dimensional elasticity problems by theoretical, numerical, or experimental methods is much more complicated than two-dimensional problems in which the state of stress is defined by three stresses. The principles of two-dimensional photoelasticity have been extended to three dimensions. However, the interpretation of the optical patterns is formidable. This comes from the variation of the magnitude and directions of stresses along the light path. It is worth mentioning that in a three-dimensional body for a constant value of the ratio C ($=d\varphi/d\delta$, $d\varphi$ is the rotation of the principal axes and $d\delta$ is the corresponding retardation) of the rotation of the directions of the principal stresses to the optical retardation, the resultant retardation through the thickness of the body is $\delta = \delta_0 (1 + 4C^2)^{0.5}$ where δ_0 is the retardation in the absence of rotation of the principal stresses [13].

Two main approaches have been developed for the solution of three-dimensional stress problems by photoelasticity: The **nondestructive** or **integrated photoelasticity** approach is based on the measurement of the integrating optical effect when a beam of light passes through the body [13, 14]. The **stress-frozen photoelasticity** approach, which is based on a property of epoxies to *freeze or lock-in the stresses and the resulting birefringent effect after the loads are removed.* We briefly present the basics of stress-frozen photoelasticity method.

Epoxies exhibit an amazing behavior when heated above a critical temperature of about 120 °C. At this temperature, the material shows a perfectly elastic behavior. No permanent deformation is left after it is unloaded. However, when the material is cooled under load up to ambient temperature the deformations and optical effects produced at temperatures above the critical temperature are retained. They are frozen or locked in.

This behavior of the material can be explained by considering that it consists of two parts or phases. A strong part and a soft part. The properties of the strong part do not change appreciably with temperature and they remain strong. The soft part at high temperatures becomes viscous and loses its load carrying capacity. At low temperatures the load is carried up by both parts. At high temperatures, almost all load is taken by the strong part, while the soft part takes only a small part of the load (since it becomes viscous) and transmits it to the strong part. When now the temperature is dropped while the load is still applied, the soft part becomes rigid again and the deformation and birefringence are locked in the strong part.

To understand this behavior consider the following analogy: The strong part of the epoxy is simulated by a network of springs and the soft part by water at elevated temperatures. The springs are immersed in a tank of water (Fig. 7.8). When a load is applied to the material at ambient temperature and it is released, then the springs will be loaded and will be returned to their unstrained positions. Consider the load is applied at elevated temperatures and is retained while the material is cooled down to ambient temperature. At ambient temperature the water freezes, while the load is still retained on the springs of the strong material. When the applied load has been released the deformation of the springs, and the resulting birefringence, is retained because the springs are blocked by the frozen water. The deformation remains fixed in the body. In the analogous case of epoxies, the resulting birefringence is locked in the material when is it loaded at elevated temperatures and cooled down to ambient temperature while the load is still retained.

The above behavior of epoxies is the basis of the stress-frozen method of photoelasticity. Since the birefringence is locked in the material, we can cut the model into slices without disturbing the birefringence. The slices are examined photoelastically in the plane or circular polariscope. The difference between the principal stresses and

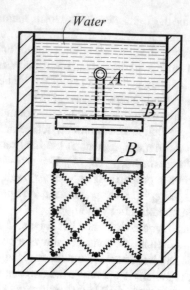

Fig. 7.8 Water–ice analogy of photoelastic materials for three-dimensional photoelasticity

their directions are determined from the isochromatic and isoclinic fringe patterns. For the case of a three-dimensional body, duplicate models are manufactured and sliced at three perpendicular directions. From the photoelastic analysis of slices of the duplicate models at three directions five out of six stresses are determined. The sixth stress is calculated from the equations of equilibrium or compatibility.

The method of integrated photoelasticity has been advanced by Aben [14] who laid down the foundations of the method. He established the equivalence theorem of three-dimensional photoelasticity, according to which *any three-dimensional photoe-lastic model is optically equivalent to a system of a birefringent plate and a rotator.* Furthermore, he introduced the use of the Faraday and Kerr effects in photoelasticity. According to the Faraday effect when a photoelastic model is inserted in a magnetic field the plane of polarization is rotated, while according to the Kerr effect when a model is placed in an electric field additional birefringence is introduced. He formulated two methods of integrated photoelasticity called magnetophotoelasticity and electrophotoelasticity based on the Faraday and Kerr effects, respectively.

7.15 Photoelastic Coatings

7.15.1 Introduction

The method of **photoelastic or birefringent coatings** allows determination of strains on the surface of opaque materials using the principles of two-dimensional photoelasticity. The method consists of cementing a coating of transparent birefringent material on the surface of the body under study and measuring the direction and the difference of the principal stresses using a reflecting polariscope. The strains are transferred from the body to the coating through the bond between them.

A major advantage of the method of photoelastic coatings is that measurements can be made of surface strains of a structural member in situ, whereas in photoelasticity measurements are made on a transparent model of the real structure. The method extends the realm of applicability of photoelasticity to opaque materials. It has many similarities with the method of strain gages for measuring strains. However, photoelastic coatings provide a full-field method, whereas strain gages measure strains at given points.

In the following, we will briefly present the transfer of stresses from body to coating, the determination of stresses, the reinforcing effect of the coatings, and the photoelastic strain gages.

7.15.2 Transfer of Stresses From Body to Coating

Consider a stressed metallic body and a thin transparent photoelastic coating perfectly bonded to the surface of the body. The surface strains of the body are transferred to the coating. The conditions of continuity of strains along the specimen/coating interface are

$$\varepsilon_1^c = \varepsilon_1^s \tag{7.47}$$

where ε_1 and ε_2 are the principal strains and the indices c and s refer to the coating and the body, respectively.

Assuming conditions of plane stress for both the body and the coating we have

$$\varepsilon_1^c = \frac{1}{E^c}\left(\sigma_1^c - v^c \sigma_2^c\right), \quad \varepsilon_2^c = \frac{1}{E^c}\left(\sigma_2^c - v^c \sigma_1^c\right)$$

$$\varepsilon_1^s = \frac{1}{E^s}\left(\sigma_1^s - V^s \sigma_2^s\right), \quad \varepsilon_2^s = \frac{1}{E^c}\left(\sigma_2^s - v^s \sigma_1^s\right) \tag{7.48}$$

From Eqs. (7.47) and (7.48) we obtain for σ_1^s and σ_2^c

$$\sigma_1^s = \frac{E^s}{E^c[1 - (\nu^s)^2]}\left[(1 - \nu^c\nu^s)\sigma_1^c - (\nu^s - \nu^c)\sigma_2^c\right]$$

$$\sigma_2^s = \frac{E^s}{E^c[1 - (\nu^s)^2]}\left[(\nu^s - \nu^c)\sigma_1^c - (1 - \nu^c\nu^s)\sigma_2^c\right] \qquad (7.49)$$

Equations (7.49) render

$$\sigma_1^s - \sigma_2^s = \frac{E^S}{E^c}\frac{1 + \nu^c}{1 + \nu^s}(\sigma_1^c - \sigma_2^c) \qquad (7.50)$$

Equation (7.50) gives the difference of the principal stresses $(\sigma_1^s - \sigma_2^s)$ in the body in terms of the difference of the principal stresses $(\sigma_1^c - \sigma_2^c)$ in the coating. $(\sigma_1^c - \sigma_2^c)$ and the directions of principal stresses can be determined using a reflection-type polariscope.

7.15.3 Determination of Stresses

The stresses in the photoelastic coating method can be determined using a reflection-type plane or circular polariscope. The arrangement of the optical elements in a reflecting polariscope is the same as that in a transmitting polariscope. The only difference between the two polariscopes is that in a transmitting polariscope the optical elements are arranged on the same axis on both sides of the body, whereas in the reflecting polariscope all elements lie on the same side of the body. Figure .7.9 shows two common types of reflecting polariscope. In Fig. 7.9a different filters as used as polarizer and analyzer and two quarter-wave plates, while in Fig. 7.9b using a beam splitter the same polarizing filter is used as polarizer and analyzer and the same quarter-wave plate serves the purpose of the two quarter-wave plates of the transmitting polariscope. In both arrangements of the polariscopes of Fig. 7.9 the light passes twice through the photoelastic coating. From Eqs. (7.50) and (7.26) we obtain

$$\sigma_1^s - \sigma_2^s = \frac{E^S}{E^c}\frac{1 + \nu^c}{1 + \nu^s}\frac{Nf_\sigma}{2d^c} \qquad (7.51)$$

where d^c is the thickness of the coating.

Equation (7.51) allows determination of $(\sigma_1^s - \sigma_2^s)$ in the body from the isochromatic fringe order N in the coating and the material constants of the specimen and the coating. The number 2 (two) in the denominator of Eq. (7.51) enters because the light passes twice through the thickness of the coating.

Equation (7.51) is based on the assumption that conditions of plane stress apply to both the body and the coating. When no assumption is made on the state of stress in the body the difference of the principal strains at the body/coating interface can be determined from Eq. (7.48) as

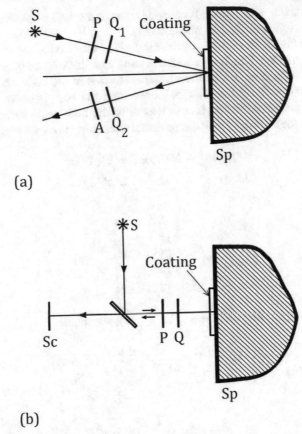

Fig. 7.9 Two arrangements of a reflecting circular polariscope used in the method of birefringent coatings. **a** Different filters are used as polarizer and analyzer with two different quarter-wave plates. **b** The same polarizing filter is used as polarizer and analyzer and the same quarter-wave plate serves the purpose of two quarter-wave plates using a beam splitter

$$\varepsilon_1^s - \varepsilon_2^s = \varepsilon_1^c - \varepsilon_2^c = \frac{1 + \nu^c}{E^c} \frac{N f_\sigma}{2d^c} \qquad (7.52)$$

Equation (7.52) allows the determination of the difference of the principal strains $(\varepsilon_1^s - \varepsilon_2^s) = (\varepsilon_1^c - \varepsilon_2^c)$ in the body/coating interface from measurement of the photoelastic fringe order N in the coating.

7.15.4 Reinforcing Effect

In the method of photoelastic coatings, the surface strains of the body are transferred to the coating by shear forces developed at the body-coating interface. This transfer results in a relaxation of the loads carried by the body. This phenomenon is called the *reinforcing effect* of photoelastic coatings.

In order to obtain an estimate of the reinforcing effect consider an infinitesimal element of area $dxdy$ of the body and the coating with thicknesses h^s and h^c, respectively. Let σ_1^s, σ_2^s and σ_1^c, σ_2^c be the principal stresses in the body and the coating respectively, and σ_1^0, σ_2^0 the principal stresses in the body without the coating. Equating the forces developed in an element of the body and coating and in the body (without the coating) along the principal stress directions we obtain

$$h^s(dy)\sigma_1^s + h^c(dy)\sigma_1^c = h^s(dy)\sigma_1^0$$
$$h^s(dx)\sigma_2^s + h^c(dx)\sigma_2^c = h^s(dy)\sigma_2^0 \tag{7.53}$$

or

$$\sigma_1^s + \frac{h^c}{h^5}\sigma_1^c = \sigma_1^0$$
$$\sigma_2^s + \frac{h^c}{h^s}\sigma_2^c = \sigma_2^0 \tag{7.54}$$

For conditions of plane stress using Eq. (7.47), we obtain

$$\varepsilon_1^0 - \varepsilon_2^0 = \left(1 + \frac{h^c}{h^s}\frac{E^c}{E^s}\frac{1+v^s}{1+v^c}\right)\left(\varepsilon_1^c - \varepsilon_2^c\right)$$

or

$$\varepsilon_1^c - \varepsilon_2^c = \frac{1}{1+C}\left(\varepsilon_1^0 - \varepsilon_2^0\right) \tag{7.55}$$

where

$$C = \frac{h^c}{h^s}\frac{E^c}{E^s}\frac{1+v^s}{1+v^c} \tag{7.56}$$

Equation (7.55) shows that the difference of the principal strains $(\varepsilon_1^c - \varepsilon_2^c)$ in the coating-body interface when the reinforcing effect of the photoelastic coating is taken into consideration is linearly related to the difference of the principal strains in the body $(\varepsilon_1^0 - \varepsilon_2^0)$ when the reinforcing effect is not considered. C is the correction factor that takes into consideration the reinforcing effect. Note that C increases as the thickness of the photoelastic coating h^c increases. Expressions for the correction factor C were obtained for many states of stress, as flexure of plates, combined plane stress and flexural loads, torsion, pressurized cylindrical tubes, etc.

Besides the reinforcing effect, the accuracy of photoelastic coatings is also affected by other factors such as the edge, strain gradient, and curvature effects, and the effect of different body-coating Poisson's ratios. These effects are not studied here.

7.15.5 *Photoelastic Strain Gages*

A photoelastic strain gage is a small piece of photoelastic material that is bonded to the surface of the body under study. The operation of photoelastic strain gages is the same as that of photoelastic coatings. Two forms of strain gages are mainly used, the uniaxial and the biaxial. The uniaxial gage has the form of a longitudinal strip and responds only to longitudinal strains parallel to its axis. Its operation is similar to strain gages. For a two-dimensional stress field, three gages are needed for the determination of the three strains at a point. The biaxial strain gage has the form of a small circular ring. It can be used to determine the directions of the principal strains and their difference. To increase the sensitivity of measuring strains by the photoelastic gages initial fringes are introduced by a previous frozen-stress cycle.

Further Readings

1. Frocht MM (1941) Photoelasticity, vol I. Wiley
2. Frocht MM (1948) Photoelasticity, vol II. Wiley
3. Coker EG, Filon LNG (1957) A treatise on photoelasticity. Cambridge University Press
4. Zandman F (1959) Photostress: principles and applications. Tatnall Measuring Systems
5. Frocht MM (ed) (1963) Photoelasticity. In: Proceedings of the international symposium held at Illinois Institute of Technology, Chicago, Illinois, Pergamon, Oct 1961
6. Durelli AJ, Riley WF (1965) Introduction to photomechanics. Prentice-Hall Inc., Boca Raton
7. Holister GS (1967) Experimental stress analysis: principles and methods. Cambridge University Press, Cambridge
8. Heywood RB (1969) Photoelasticity for designers. Pergamon Press
9. Leven MM (ed) (1969) The selected scientific papers of MM Frocht. Pergamon
10. Brčić V (1970) Photoelasticity in theory and practice. Springer, Berlin
11. Kuske A, Robertson G (1974) Photoelastic stress analysis. Wiley
12. Brčić V (1975) Application of holography and hologram interferometry to photoelasticity. Springer, Berlin
13. Theocaris PS, Gdoutos EE (1979) Matrix theory of photoelasticity. Springer
14. Aben H (1979) Integrated photoelasticity. McGraw-Hill
15. Paipetis SA, Holister GS (eds) (1985) Photoelasticity in engineering practice. Elsevier, Amsterdam
16. Gdoutos EE (1985) Photoelastic study of crack problems. In: Paipetis AS, Hollister GS (eds) Photoelasticity in engineering practice. Elsevier, Barking, pp 181–204
17. Gdoutos EE (1985) Photoelastic analysis of composite materials with stress concentrations. In: Paipetis AS, Hollister GS (eds) Photoelasticity in engineering practice. Elsevier, Barking, pp 157–179
18. Nisida M, Kawata K (eds) (1986) Photoelasticity. In: Proceedings of the international symposium on photoelasticity, Tokyo. Springer, Berlin
19. Lagarde A (1987) (ed) Static and dynamic photoelasticity and caustics. Recent developments. Springer, Berlin
20. Dally JW, Riley WF (1991) Experimental stress analysis. McGraw-Hill, pp 366–374, 424–531
21. Burger CP (1993) Photoelasticity. In: Kobayashi AS (ed) Handbook of experimental mechanics, 2nd edn. Society for Experimental Mechanics, pp 165–296
22. Aben H, Guillemet C (1993) Photoelasticity of glass. Springer
23. Orr JF, Finlay JB (1997) Photoelastic stress analysis. In: Orr JF, Shelton JC (eds) Optical measurement methods in biomechanics, Chapman & Hall, pp 1–16

24. Cloud GL (1998) Optical methods in engineering analysis. Cambridge University Press, Cambridge, pp 57–144
25. Chen TF (ed) (1999) Selected papers on photoelasticity. SPIE Press
26. Khan AS, Wang X (2001) Strain measurements and stress analysis. Prentice Hall, Boca Raton, pp 94–173
27. Ramesh K (2000) Digital photoelasticity. Springer
28. Ramesh K (2008) Photoelasticity. In: Sharpe WN (ed) Handbook of experimental solid mechanics. Springer, pp 701–742
29. Razumovsky IA (2011) Interference-optical methods of solid mechanics. Springer, Berlin, pp 1–35
30. Sciammarella CA, Sciammarella FM (2012) Experimental mechanics of solids. Wiley, New York, pp 285–386
31. Shukla A, Dally JW (2014) Experimental solid mechanics, 2nd edn. College House Enterprise, LLC, pp 283–362
32. Micro Measurements (2015) Calibration of photostress coatings. Tech Note 701
33. Micro Measurements (2015) How to select photoelastic coatings. Tech Note 704
34. Post D (1966) Fringe multiplication in three-dimensional photoelasticity. J Strain Anal 1(5):380–388
35. Post D (1970) Photoelastic-fringe multiplication: for tenfold increase in sensitivity. Exp Mech 10(8):305–312

Chapter 8
Interferometry

8.1 Introduction

Interferometric methods of stress analysis are based on the phenomenon of interference of light. They use interferometers to measure the optical retardations of bodies due to loading, which are directly related to the stresses. When combined with the difference of the principal stresses obtained from isochromatic patterns of photoelasticity the individual principal stresses are determined.

In this chapter, we present the Mach–Zehnder, the Michelson, and the Fizeau-type interferometers for stress analysis. The optical patterns obtained by these interferometers are studied using the Jones calculus. At the end of the chapter, we present a generic interpretation of the optical patterns obtained by all interferometers.

8.2 Interferometers

Interferometers are devices that divide a light ray into two rays and bring them together to produce interference. They have applications in optics, astronomy, mechanics, etc. In this chapter, interferometers are used for measuring stresses. One ray called *active ray* passes through the model, while the other ray called *reference ray* does not pass through the model. The optical path of the active ray changes when the model is loaded, while the optical path of the reference ray does not change. The active and reference rays interfere. The interference pattern provides information about the variation of the optical path of the active ray that is related to the stresses in the model. Figure 8.1 shows schematic diagrams of eight interferometers used in experimental mechanics.

Figure 8.1a shows the **Mach–Zehnder** interferometer. The optical paths of the active and reference rays are equal. When the model is inserted in the path of the active ray the optical paths of the active and reference rays differ. To compensate for this difference a dummy model is inserted in the path of the reference ray. When

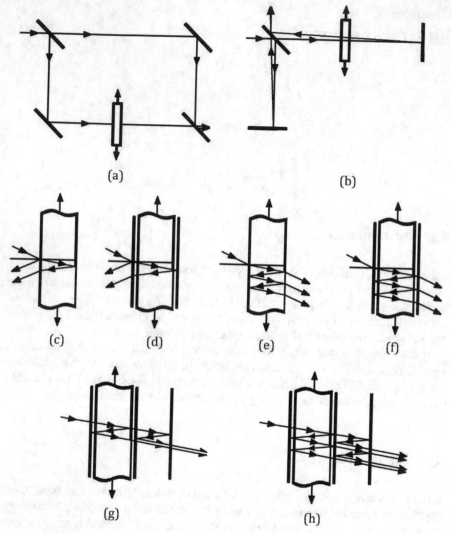

Fig. 8.1 Schematic diagrams of interferometers used in experimental mechanics

the model has loaded the optical paths of the two rays differ by the stress-optical retardation of the model due to loading.

In the Michelson interferometer (Fig. 8.1b) the light passes twice through the model. A dummy model is introduced in the reference ray. The change of the optical path is twice that of the Mach–Zehnder interferometer. The difference of the optical paths in both Michelson and Mach–Zehnder interferometers is usually not many times greater than the wavelength of light. Thus, a high-quality interference pattern is obtained. To further minimize the difference of the optical paths between the active and reference rays and to enhance the quality of the interference fringe pattern the interferometer is inserted in an oil bath of the same index of refraction as the model.

In Fig. 8.1c no external optical elements are used. The optical pattern is formed by the interference of the rays reflected from the front and the rear faces of the model. The reference ray is reflected from the front face and does not pass through the model, while the active ray passes twice through the thickness of the model.

In Fig. 8.1d the mechanism of formation of fringes is the same as in Fig. 8.1c with the difference that the incident light is not reflected from the faces of the model, but from two mirrors placed one in front and the other behind the model. The reflectivity of the two mirrors is approximately the same so that good quality interference fringes are obtained.

In the Fabry–Perot interferometric systems of Fig. 8.1e, f the reference ray passes once through the thickness of the model. Of the two active rays, the first pass three times through the thickness of the model and interferes with the reference ray, while the second passes five times. In the system of Fig. 8.1e no external optical element is used, whereas in the system of Fig. 8.1f two mirrors are placed, one in front and the other behind the model. In the series interferometers of Fig. 8.1 g, h the reference ray passes once, whereas the active ray passes three or more times through the thickness of the model.

In the case of Fig. 10.4c–f the optical-path difference of the interfering rays is large compared with the wavelength of light. Thus, a light source of high coherence is needed. In the interferometric systems of Fig. 10.4g, h the distance between the two mirrors behind the model is arranged so that the two interfering rays have approximately equal optical-path lengths.

8.3 Analysis of Interferometers

8.3.1 Introduction

We analyze the Mach–Zehnder (Fig. 8.1a), the Michelson (Fig. 8.1b), and the Fizeau-type (Fig. 8.1c) interferometers. The analysis of the optical patterns of the other interferometers of Fig. 8.1 can be obtained from the analysis of the optical patterns of these three interferometers.

8.3.2 The Mach–Zehnder Interferometer

Figure 8.2 shows the arrangement of the optical elements in the Mach–Zehnder interferometer. A monochromatic light emitted from a source S passes through a pinhohe P placed at the focus of a lens L_1. The collimated beam exiting the lens impinges on a beam splitter or half-mirror H_1 at an angle 45°. Two perpendicular

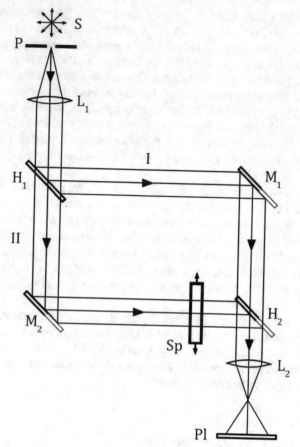

Fig. 8.2 Arrangement of the optical elements and the specimen in the Mach–Zehnder interferometer

beams *I* and *II* are obtained. Beam *I* is reflected from mirror M_1 and passes through a beam splitter H_2. This is the reference beam. Beam *II* is reflected from mirror M_2, passes through the transparent model *Sp* and impinges on the beam splitter H_2. This is the active ray. Beams *I* and *II* interfere. The interference pattern is received on a photographic plate *Pl*. Usually, a dummy model of the same thickness as the model under study is inserted in the optical path of the reference beam *I*.

The optical paths of the active and reference rays when the model is unloaded are equal and no interference pattern is formed. However, due to variation of the specimen thickness of the order of the wavelength of light interference fringes on the ground glass of the camera may appear. When the specimen is loaded the optical paths Δs_1 and Δs_2 along the principal stress directions given by Eqs. (6.34) and (6.35) change. The Jones vector *J* of the specimen referred to its principal axes is

$$J = \begin{bmatrix} e^{-i\delta_1} & 0 \\ 0 & e^{-i\delta_2} \end{bmatrix} \tag{8.1}$$

where

$$\delta_1 = \frac{2\pi}{\lambda} \Delta s_1 = \frac{2\pi}{\lambda}(a_t\sigma_1 + b_t\sigma_2)d$$

$$\delta_2 = \frac{2\pi}{\lambda} \Delta s_2 = \frac{2\pi}{\lambda}(a_t\sigma_2 + b_t\sigma_1)d \tag{8.2}$$

λ is the wavelength of the light, a_t and b_t are the stress-optical constants for light rays transmitted through the model, σ_1 and σ_2 are the principal stresses and d is the thickness of the model.

Consider now that a unit-intensity right-handed circularly polarized light beam emerges from lens L_1. The Jones vector a' of the light beam is given by

$$a' = \frac{1}{\sqrt{2}} \begin{bmatrix} -i \\ 1 \end{bmatrix} \tag{8.3}$$

The light beam a'' that emerges from the model is

$$a'' = \frac{1}{\sqrt{2}} \begin{bmatrix} e^{-i\delta_1} & 0 \\ 0 & e^{-i\delta_2} \end{bmatrix} \begin{bmatrix} -i \\ 1 \end{bmatrix} = \frac{1}{\sqrt{2}} \begin{bmatrix} -ie^{-i\delta_1} \\ e^{-i\delta_2} \end{bmatrix} \tag{8.4}$$

Equations (8.3) and (8.4) represent the Jones vectors of the reference and active beams. The two beams interfere. The Jones vector a of their interference is

$$a = a' + a'' = \frac{1}{\sqrt{2}} \begin{bmatrix} -i(e^{-i\delta_1} + 1) \\ e^{-i\delta_2} + 1 \end{bmatrix} \tag{8.5}$$

The intensity I of the beam with Jones vector a is

$$I = \tilde{a}a = \frac{1}{2}\left[i(e^{i\delta_1} + 1) \ \ e^{i\delta_2} + 1 \right] \begin{bmatrix} -i(e^{-i\delta_1} + 1) \\ e^{-i\delta_2} + 1 \end{bmatrix} \tag{8.6}$$

or

$$I = 2 + \cos\delta_1 + \cos\delta_2 \tag{8.7}$$

Using Eq. (8.2) we obtain

$$I = 2\left\{ 1 + \cos\left[\frac{2\pi}{\lambda} \frac{a_t + b_t}{2}(\sigma_1 + \sigma_2)d \right] \cos\left[\frac{2\pi}{\lambda} \frac{a_t - b_t}{2}(\sigma_1 - \sigma_2)d \right] \right\} \tag{8.8}$$

Equation (8.8) shows that the intensity of the optical pattern obtained in the Mach–Zehnder interferometer depends on the values of the sum $(\sigma_1 + \sigma_2)$ and the difference $(\sigma_1 - \sigma_2)$ of the principal stresses. Therefore, *the interference pattern contains the superimposed families of isochromatics and isopachics.* This result was obtained independently by Nisida and Saito [2].

8.3.3   The Michelson Interferometer

Figure 8.3 shows the arrangement of the optical elements in the Michelson interferometer. A monochromatic light emitted by a source S passes through a pinhohe P placed at the focus of a lens L_1. The obtained collimated beam exiting the lens impinges on a beam splitter or half-mirror H at an angle 45°. It is divided into two beams that follow paths I and II. Beam I and II are reflected by mirrors M_1 and M_2 which are placed at equal distances from the beam splitter H. After passing through the beam splitter H the two beams interfere. The interference pattern is received on the photographic plate Pl of a lens L_2. The model is inserted in the optical path of either beam I or II (in beam II in Fig. 8.3). This is the active beam. Beam I is the reference beam. To obtain equal optical path lengths a dummy specimen is inserted in beam I. The interferometer can also be immersed in a tank with oil of the same

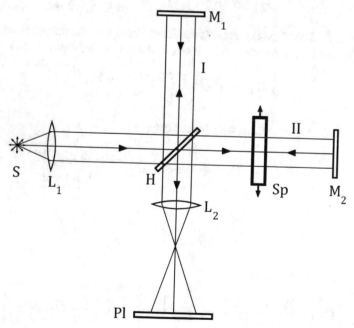

Fig. 8.3 Arrangement of the optical elements and the specimen in the Michelson interferometer

index of refraction as that of the specimen. When the specimen is unloaded an interference pattern appears which depicts the thickness irregularities of the specimen of the order of the wavelength of light.

The model is traversed twice by the active beam. Let J be the Jones matrix of the model traversed by the light beam in one direction. If the model is traversed in the opposite direction the Jones matrix is given by its transpose. Because the Jones matrix J is diagonal, its transpose is equal to the matrix itself. Thus, the Jones matrix J_I of the specimen which is traversed twice by the active beam II is

$$J_1 = J^2 = \begin{bmatrix} e^{-i\delta_1} & 0 \\ 0 & e^{-i\delta_2} \end{bmatrix} \begin{bmatrix} e^{-i\delta_1} & 0 \\ 0 & e^{-i\delta_2} \end{bmatrix} = \begin{bmatrix} e^{-2i\delta_1} & 0 \\ 0 & e^{-2i\delta_2} \end{bmatrix} \tag{8.9}$$

From Eqs. (8.1) and (8.9) it is concluded that the angles entering in the Jones matrix of the model in the Michelson interferometer are twice the angles in the Jones matrix of the model in the Mach–Zehnder interferometer. Thus, *the optical effect obtained by the Michelson interferometer is the same as that of the Mach–Zehnder interferometer with the only difference that the angles in Eq. (8.8)* must be doubled.

8.3.4 The Fizeau-Type Interferometer

In the Fizeau-type interferometer (Fig. 8.1c) no external optical elements are used. The reference ray is reflected from the front face of the model, while the active ray passes twice through the thickness of the model and is reflected from its rear face. The optical pattern is formed by the interference of the active and reference rays.

The Jones matrix J of the specimen is

$$J = \begin{bmatrix} e^{-i\delta_1} & 0 \\ 0 & e^{-i\delta_2} \end{bmatrix} \tag{8.10}$$

where

$$\delta_1 = \frac{2\pi}{\lambda} \Delta s_{r_1} = \frac{2\pi}{\lambda} 2(a_r \sigma_1 + b_r \sigma_2)d$$

$$\delta_2 = \frac{2\pi}{\lambda} \Delta s_2 = \frac{2\pi}{\lambda} 2(a_r \sigma_2 + b_r \sigma_1)d \tag{8.11}$$

The stress-optical constants a_r and b_r are related to the light rays reflected from the rear face of the model. They are given by Eq. (6.41a).

Consider that a unit-intensity right-handed circularly polarized light illuminates the model. The Jones vector of the active beam emerging from the model is obtained from Eq. (8.4) as

$$a'' = \frac{1}{\sqrt{2}} \begin{bmatrix} -ie^{-i\delta_1} \\ e^{-i\delta_2} \end{bmatrix} \tag{8.12}$$

where δ_1 and δ_2 are given from Eq. (8.11).

The reference ray is reflected from the front face of the model. When the model is loaded it undergoes a change of the optical path equal to $-(\Delta d)n_0$, where (Δd) is the change of the thickness and n_0 is the index of refraction of the medium that surrounds the model.

The Jones matrix of the model is given by Eq. (8.10) as

$$J = \begin{bmatrix} e^{-i\delta_1'} & 0 \\ 0 & e^{-i\delta_2'} \end{bmatrix} \tag{8.13}$$

where under conditions of plane stress

$$\delta_1' = \delta_2' = -\frac{2\pi}{\lambda}(\Delta d)n_0 = \frac{2\pi}{\lambda}\frac{v(\sigma_1+\sigma_2)d}{E}n_0 \tag{8.14}$$

The Jones vector of the reference ray is given by Eq. (8.12) as

$$a''' = \frac{1}{\sqrt{2}} \begin{bmatrix} -ie^{-i\delta_1'} \\ e^{-i\delta_2'} \end{bmatrix} \tag{8.15}$$

The active and reference rays interfere. The Jones vector a of the light obtained from the interference of the active and reference rays is

$$a = a'' + a''' = \frac{1}{\sqrt{2}} \begin{bmatrix} -ie^{-i\delta_1} \\ e^{-i\delta_2} \end{bmatrix} + \frac{1}{\sqrt{2}}\frac{1}{\sqrt{2}} \begin{bmatrix} -i(e^{-i\delta_1}+e^{-i\delta_1'}) \\ e^{-i\delta_2}+e^{-i\delta_2'} \end{bmatrix} \tag{8.16}$$

The intensity I of light with Jones vector a is

$$I = \tilde{a}a = \frac{1}{2}\left[i\left(e^{i\delta_1}+e^{i\delta_1'}\right)e^{i\delta_2}+e^{i\delta_2'}\right]\begin{bmatrix} -i\left(e^{-i\delta_1}+e^{-i\delta_1''}\right) \\ e^{-i\delta_2}+e^{-i\delta_2'} \end{bmatrix} \tag{8.17}$$

or

$$I = 2\left\{1+\cos\left[\frac{2\pi}{\lambda}(a^*+b^*)(\sigma_1+\sigma_2)d\right]\cos\left[\frac{2\pi}{\lambda}(a^*-b^*)(\sigma_1-\sigma_2)d\right]\right\} \tag{8.18}$$

where

$$a^* = \frac{1}{E}[b_1 - 2vb_2 - vn] = A - \frac{v}{E}n$$

$$b^* = \frac{1}{E}[b_1 - \nu(b_1 + b_2) - \nu n] = B - \frac{\nu}{E}n \qquad (8.19)$$

Equation (8.19) indicates that *the optical effect obtained in the Fizeau-type interferometer is the same as that of the Mach–Zehnder and Michelson interferometers with the only difference that the constants a* and b* are expressed by Eq. (8.19).*

8.3.5 Other Interferometers

The analysis of the above three interferometers forms the basis for the analysis of the other interferometers of Fig. 8.1.

The Fizeau-type interferometer of Fig. 8.1d is equivalent to the Michelson interferometer. The light traverses the specimen twice and it is reflected from mirrors, not the faces of the specimen.

The interferometer of Fig. 8.1e is equivalent to that Fig. 8.1c. Indeed, only the first two rays that emerge from the rear face of the specimen contribute to the formation of the interference pattern. The intensity of the other rays is insignificant (see Sect. 6.4).

The interferometer of Fig. 8.1f is equivalent to that of Fig. 8.1d.

Finally, the interferometers of Fig. 8.1 g, h are equivalent to the interferometers of Fig. 8.1d, f, respectively.

8.3.6 A Generic Analysis of Interferometers

From the analysis of all interferometers of Fig. 8.1 *it is shown that the superimposed families of isochromatics and isopachics are obtained.* This result can be obtained by resorting to physical concepts only. All interferometers divide the incident light into reference and active rays. In some interferometers, the active ray passes through the model β times more than the reference ray. Thus the stress-optical retardation along the principal stress directions σ_1 and σ_2 between the active and reference rays is $\beta(\Delta s_1)$ and $\beta(\Delta s_2)$, where (Δs_1) and (Δs_2) are the stress-optical retardation when light passes once through the model thickness. In all interferometers, the fringe orders N_1 and N_2 for the patterns along the directions of the principal stresses are given by

$$N_{1,2} = \frac{\beta(\Delta s_{1,2})}{\lambda} \qquad (8.20)$$

Introducing the values of the stress-optical retardations $\Delta s_{1,2}$ from Eqs. (6.34), (6.35) into Eq. (8.20) we obtain

$$N_1 = (A\sigma_1 + B\sigma_2)$$

$$N_2 = (B\sigma_1 + A\sigma_2) \tag{8.21}$$

where

$$A = \frac{\beta a}{\lambda}, B = \frac{\beta b}{\lambda} \tag{8.22}$$

Equations (8.21) indicate that the optical patterns in all interferometers consist of two overlapping families of curves which express the quantities $(A\sigma_1 + B\sigma_2)$ and $(B\sigma_1 + A\sigma_2)$. These two families of curves $N_1 = N_1(x,y)$ and $N_2 = N_2(x,y)$, according to the moiré effect, combine to form two new families of curves given by

$$M_1 = N_1 - N_2$$
$$M_2 = N_1 + N_2 \tag{8.23}$$

or using Eq. (8.21)

$$M_1 = (A - B)(\sigma_1 - \sigma_2)d$$
$$M_2 = (A + B)(\sigma_1 + \sigma_2)d \tag{8.24}$$

Equation (8.24) shows that the interferometric fringe patterns consist of the super-imposed families of isochromatics and isopachics. As it was indicated in Sect. 3.4, from the two moiré patterns, the subtractive and the additive, only one pattern is visible. The two patterns are separated by the *commutation moiré boundary* given by Eq. (3.15), as

$$\psi(x, y) = \frac{\partial N_1}{\partial x}\frac{\partial N_2}{\partial x} + \frac{\partial N_1}{\partial y}\frac{\partial N_2}{\partial y} = 0 \tag{8.25}$$

or using Eq. (8.21) by

$$\psi(x, y) = AB\left[\left(\frac{\partial\sigma_1}{\partial x}\right)^2 + \left(\frac{\partial\sigma_1}{\partial y}\right)^2 + \left(\frac{\partial\sigma_2}{\partial x}\right)^2 + \left(\frac{\partial\sigma_2}{\partial y}\right)^2\right]d$$
$$+ (A^2 + B^2)\left(\frac{\partial\sigma_1}{\partial x}\frac{\partial\sigma_2}{\partial x} + \frac{\partial\sigma_1}{\partial y}\frac{\partial\sigma_2}{\partial y}\right)d \tag{8.26}$$

When $\psi(x,y) > 0$ the subtractive pattern defined by $N = N_1 - N_2$ is effective, whereas when $\psi(x,y) < 0$ the additive pattern defined by $N = N_1 + N_2$ is effective.

From the foregoing discussion, it is concluded that four superimposed patterns $(A\sigma_1 + B\sigma_2)$, $(B\sigma_1 + A\sigma_2)$, $(\sigma_1 + \sigma_2)$, $(\sigma_1 - \sigma_2)$ are formed in all interferometers. Which of the four families of curves appear in an interferometric pattern is case specific and depends on many factors. It will not be discussed here. However, it can be pointed out that for optically inert materials the change of the optical paths along the principal stress directions are equal and only isopachic fringes are obtained.

It should be noted that the initial interferometric pattern produced by small variations of the model thickness can be subtracted from the final pattern by superimposing the two patterns. The new pattern obtained, according to the moiré effect, represents the difference between the two patterns and thus the initial pattern is eliminated.

Further Readings

1. Theocaris PS, Gdoutos EE (1979) Matrix theory of photoelasticity. Springer, Berlin, pp 184–207
2. Nisida M, Saito H (1964) A new interferometric method of two-dimensional stress analysis. Exp Mech 4:366–376

Chapter 9
Holography

9.1 Introduction

Holography is an optical method for recording both the amplitude and the phase of a wavefront. Its name comes from the Greek words "*holo*" which means "*whole*" and "*graphy*" which means "*recording*". The term "wavefront reconstruction" is also used. Before holography the only method of recording and retaining as a permanent record the picture of an object was photography. In photography, the wavefront emitted by an object is transformed by a lens and impinges on a photosensitive plate that responds to the intensity of light. Thus, only the amplitude, not the phase, of the wave can be recorded. *Holography is based on the phenomena of interference and diffraction of light.*

Holography is a two-step lensless imaging process of wavefront reconstruction. It was invented by Gabor in 1948. Its evolution was helped with the invention of Laser (acronym for "light amplification by stimulated emission of radiation") in 1960.

This chapter presents the basic principles of holography and holographic interferometry. We develop the equations of the recording and reconstruction processes using the Jones calculus and a generic explanation of both holography and holographic interferometry. For more information on holography and its use in stress analysis, the reader is referred to the excellent book in the "References" at the end of this chapter.

9.2 Recording and Reconstruction Processes

Holography is a two-step process for the complete recording of a wavefront emitted by an object. In the first step, called the **recording process**, the object wave interferes with a known, usually plane, reference wave. The resulting interference pattern is stored in a recording medium, usually a photographic plate with a high-resolution

© The Author(s), under exclusive license to Springer Nature Switzerland AG 2022
E. E. Gdoutos, *Experimental Mechanics*, Solid Mechanics and Its Applications 269,
https://doi.org/10.1007/978-3-030-89466-5_9

emulsion. The amplitude and the phase of the object wave are converted into intensity variations of the interference pattern, which, after development, is called the **hologram**. In the second step, called the **reconstructing process**, the hologram is illuminated by the reference wave and the resulting diffraction pattern has the complete characteristics, that is, the amplitude and the phase, of the object wave.

We will present the equations of the recording and reconstruction processes using the Jones calculus.

a. *Recording process*. Let a_1 and a_2 be the Jones vectors of the reference beam *one* and the object beam *two*, respectively (Fig. 9.1). They are given by

$$a_1 = \begin{bmatrix} A_{x_1} e^{i\delta_{x_1}} \\ A_{y_1} e^{i\delta_{y_1}} \end{bmatrix} \quad a_2 = \begin{bmatrix} A_{x_2} e^{i\delta_{x_2}} \\ A_{y_2} e^{i\delta_{y_2}} \end{bmatrix} \tag{9.1}$$

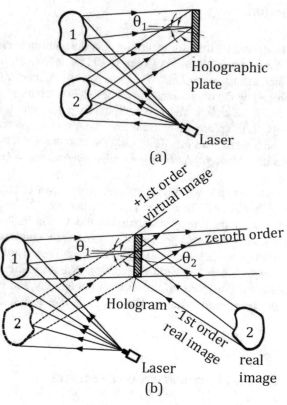

Fig. 9.1 Formation of hologram by illuminating a holographic plate with the wavefronts of objects *1* and *2* during the recording process (**a**). Reconstruction process of wavefront of object 2 by illuminating the hologram with the wavefront of object 1 (**b**). Two images of object 2 are formed by the + 1 and −1 diffraction orders. The virtual image is in the initial position of object 2 and the real image is at a symmetrical position of object 2 with respect to the hologram

Consider that waves 1 and 2 are incident on the recording plate at angles θ_1 and θ_2, respectively. Then, waves 1 and 2 at the recording plate are

$$a_1' = a_1 e^{-ikx \sin \theta_1}, \quad a_2' = a_2 e^{-ikx \sin \theta_2} \tag{9.2}$$

The above two waves interfere. The Jones vector a' of the combined wave is

$$a' = a_1' + a_2' \tag{9.3}$$

The intensity of light that corresponds to the Jones vector a' is

$$I = \tilde{a}' a' = \tilde{a}_1' a_1' + \tilde{a}_2' a_2' + \tilde{a}_1' a_2' + \tilde{a}_2' a_1' \tag{9.4}$$

where $\tilde{a}'$ is the **Hermitian** conjugate vector of vector a' (a Hermitian matrix is a complex square matrix that is equal to its own conjugate transpose—that is, the element in the ith row and jth column is equal to the complex conjugate of the element in the jth row and ith column).

The amplitude transmittance τ of the hologram is

$$\tau = kI = k\left(\tilde{a}_1' a_1' + \tilde{a}_2' a_2' + \tilde{a}_1' a_2' + \tilde{a}_2' a_1'\right) \tag{9.5}$$

where k is a constant.

b. *Reconstruction process.* In the reconstruction process, the hologram is illuminated by the reference wave a_1'. The Jones vector a_τ of the light transmitted through the hologram is

$$a_\tau = kIa_1' = ka_1'\left(\tilde{a}_1' a_1' + \tilde{a}_2' a_2' + \tilde{a}_1' a_2' + \tilde{a}_2' a_1'\right) \tag{9.6}$$

Using Eq. (9.2) we obtain the intensity I of the transmitted light

$$I = k\left[\left(A_{x_1}^2 + A_{y_1}^2\right) + \left(A_{x_2}^2 + A_{y_2}^2\right)\right] a_1 \, e^{-ikx \sin \theta_1} \, ka_1 \tilde{a}_1 a_2 e^{-ikx \sin \theta_2}$$
$$+ ka_1 \tilde{a}_2 a_1 e^{-ikx(2 \sin \theta_1 - \sin \theta_2)} \tag{9.7}$$

Let us now examine the three terms of Eq. (9.7).

The *first term* of Eq. (9.7) is the product of the Jones vector of the reference beam a_1' and the sum of the intensities of the reference and object waves, which can be assumed approximately constant all over the field. Therefore, *this term represents the reconstructed image of the reference wave.* It corresponds to the zero-order diffraction pattern from the hologram.

The *second term* of Eq. (9.7) is proportional to $a_1 \, \tilde{a}_1 \, a_2$ which using Eq. (9.1) is written as

$$a_1 \tilde{a}_1 a_2 = A_{x_1}^2 \begin{bmatrix} A_{x_2} e^{i\delta_{x_2}} \\ A_{y_2} e^{i\delta_{y_2}} \end{bmatrix} + \left(A_{y_1}^2 - A_{x_1}^2\right) \begin{bmatrix} 0 \\ A_{y_2} e^{i\delta_{y_2}} \end{bmatrix} + A_{x_1} A_{y_1} \begin{bmatrix} A_{y_2} e^{i\left(\delta_{y_2} - \delta_{y_1} + \delta_{x_1}\right)} \\ A_{x_2} e^{i\left(\delta_{x_2} - \delta_{x_1} + \delta_{y_1}\right)} \end{bmatrix}$$

$$(9.8)$$

The first term of the above expression is proportional to the object wave a_2. Therefore, *this term represents the reconstructed image by the hologram of the object wave.*

The second and third terms of Eq. (9.8) represent two waves focused in the same direction as that of a_2. The effect of these waves is to add some vagueness to the image of the object. Elimination of the second term can be achieved by using a circularly polarized light as a reference wave.

The *third term* of Eq. (9.7) is proportional to $a_1 \tilde{a}_2 a_1$. For small angles θ_1, θ_2 ($\sin\theta_{1,2} = \theta_{1,2}$) it represents a light beam that propagates in a direction that makes an angle $(\theta_1 - \theta_2)$ with the reference beam. By arguments analogous to those of the second term it can be proven that this term contains the conjugate vector $\tilde{a}_2$ of vector a_2. Therefore, *this term represents the reconstructed virtual image of the object wave.*

From the above discussion, it is shown that *by illuminating the hologram with the reference wave we obtain three waves: the reference wave, the real object wave, and the virtual object wave.* The reconstruction process is shown in Fig. 9.1b. Note that the *virtual image is formed in the original location of the object.*

A simplified explanation of the holographic process is based on the equivalence principle between interference and diffraction developed in Sect. 2.10.4. The reference and object waves interfere. Their interference pattern is recorded on the holographic plate which after development forms the hologram. *The hologram is a diffraction grating. When it is illuminated by the reference wave its diffraction pattern gives: the reference wave itself as zero-order diffraction and the real and imaginary images of the object as the two (real and virtual) first-order diffractions.*

9.3   Holographic Interferometry

9.3.1   Introduction

Holographic interferometry provides an elegant and powerful application of holography. It permits measurement of displacements and stresses without requiring high-quality optical elements, as in classical interferometry. Holographic interferometry is based on the fundamental property of the hologram to store and reconstruct a given wave. The wave is stored in the hologram, it is reconstructed and then interferes with another wave. Also, on the same hologram, various waves can be stored, and they can be reconstructed and interfere. These two principles constitute the pillars of the two main techniques of holographic interferometry, namely, the **real-time** and **double-exposure** technique. In the following, we will present these two techniques.

9.3.2 *Real-Time Holographic Interferometry*

According to the wavefront reconstruction by the two-step process of holography, the interference of the object and reference wave is recorded in the hologram. When the hologram is illuminated by the reference wave the object wave is reconstructed. Let us assume that in the reconstruction process the shape of the object is changed, e.g., due to the application of loads. The hologram is illuminated by the reference wave and the new form (deformed) of the object wave. Illumination of the hologram with the reference wave reconstructs the original object wave, whereas illumination of the hologram with the new form of the object wave gives as a zero-order diffraction pattern the new form of the object wave. The two forms of the object wave that correspond to the unloaded and the loaded shape of the object interfere. The resulting interferogram consists of fringes that are the loci of equal displacement of the body.

We will present the equations of the recording and reconstruction processes using the Jones calculus.

a.   *Recording process.* Let us consider the simple case when both the reference, a_1 and the object, a_2, waves are one-dimensional. Their Jones vectors are

$$a_1 = \begin{bmatrix} A_{x_1} e^{i\frac{2\pi}{\lambda}\theta_x x} \\ 0 \end{bmatrix}, \quad a_2 = \begin{bmatrix} A_{x_2} e^{i\delta_x} \\ 0 \end{bmatrix} \tag{9.9}$$

where the reference wave impinges at an angle θ_x with respect to the x-axis.

The intensity of the transmitted light from the hologram is

$$I = k(\tilde{a}_1 + \tilde{a}_2)(a_1 + a_2) = k\left\{ A_{x_1}^2 + A_{x_2}^2 + A_{x_1} A_{x_2} \left[e^{i\left(\delta_x - \frac{2\pi}{\lambda}\theta_x x\right)} + e^{-i\left(\delta_x - \frac{2\pi}{\lambda}\theta_x x\right)} \right] \right\} \tag{9.10}$$

b.   *Reconstruction process.* Let us now consider that the object has been deformed. The Jones a'_2 vector of the deformed object is

$$a'_2 = \begin{bmatrix} A'_{x_2} e^{i\delta'_x} \\ 0 \end{bmatrix} \tag{9.11}$$

If the hologram is illuminated by both the reference, a_1, and deformed object, a'_2, waves the wave transmitted by the hologram is

$$a_T \begin{bmatrix} A_{x_T} e^{i\delta_{x_T}} \\ 0 \end{bmatrix} \tag{9.12}$$

where

$$A_{x_T} e^{i\delta_{x_T}} = I \left(A_{x_1} e^{i\frac{2\pi}{\lambda}\theta_x x} + A'_{x_2} e^{i\delta'_x} \right)$$

$$= k \Big[A'_{x_2} A^2_{x_2} e^{i\delta'_x} + A'_{x_2} A^2_{x_1} e^{i\delta'_x} + A'_{x_2} A_{x_1} A_{x_2} e^{i\left(\delta_x + \delta'_x - \frac{2\pi}{\lambda}\theta_x x\right)}$$

$$+ A'_{x_2} A_{x_1} A_{x_2} e^{-i\left(\delta_x - \delta'_x - \frac{2\pi}{\lambda}\theta_x x\right)} + A_{x_1} A^2_{x_2} e^{i\frac{2\pi}{\lambda}\theta_x x} + A^3_{x_1} e^{i\frac{2\pi}{\lambda}\theta_x x}$$

$$+ A_{x_2} A^2_{x_1} e^{i\delta_x} + A^2_{x_1} A_{x_2} e^{-i\left(\delta_x - \frac{4\pi}{\lambda}\theta_x x\right)} \Big] \tag{9.13}$$

Equation (9.13) shows that the light transmitted by the hologram consists of many waves that travel in several directions. Of all these waves, those transmitted in approximately the direction of the original object wave are represented by the first two and the seventh terms of the right-hand side of Eq. (9.13). Thus Eq. (9.13) may be simplified as

$$A_{x_T} e^{i\delta_{x_T}} = k \Big[A'_{x_2} \left(A^2_{x_1} + A^2_{x_2} \right) e^{i\delta'_x} + A_{x_2} A^2_{x_1} e^{i\delta_x} \Big] \tag{9.14}$$

The intensity $I\prime$ of light corresponding to this wave is given by

$$I' = k^2 \Big[\left(A^2_{x_1} + A^2_{x_2} \right)^2 A'^2_{x_2} + A^4_{x_1} A^2_{x_2} + 2 A^2_{x_1} A_{x_2} \left(A^2_{x_1} + A^2_{x_2} \right) A'_{x_2} \cos\left(\delta_x - \delta'_x \right) \Big] \tag{9.15}$$

Equation (9.15) indicates that interference fringes are obtained when

$$\delta_x - \delta'_x = m \frac{\pi}{2} \tag{9.16}$$

where $m = (2n + 1)$ for dark fringes and $m = 2n$ for bright fringes (n is an integer).

Equation (9.16) shows that *the fringes obtained in holographic interferometry represent the loci of equal optical-path difference between the original and the deformed body.* Thus, the displacements of the body are determined.

In the method of real-time holographic interferometry, the hologram of the unloaded body is first obtained. The hologram is then illuminated by the reference beam and the beam that corresponds to a loaded state of the body. The resulting interferogram of the two beams that correspond to the unloaded and loaded body represents the loci of the displacements. Thus, for each state of the body, an interferogram is obtained instantaneously. Hence the name *real-time holographic interferometry.*

Experimental difficulties arise because the hologram is made at one time and it is illuminated by the new form of the object beam that corresponds to the loaded body at another time. All optical elements including the hologram and the object must be at exactly the same positions at both times. Another source of error is that the emulsion of the hologram may shrink during development and drying and may reconstruct a distorted initial object wave. The method is characterized by large flexibility in obtaining a series of interferograms of the object corresponding to its various forms during deformation.

c. *Generic interpretation.* The resulting optical effect in real-time holographic interferometry can be explained by physical arguments. A hologram is first obtained from the reference beam and the object beam corresponding to the unloaded body. The hologram is then illuminated by the reference beam and the beam from the loaded body. By illuminating the hologram with the reference beam the image of the unloaded body is reconstructed. By illuminating the hologram with the light beam from the loaded body the image of the loaded body is obtained as first-order diffraction. The images of the unloaded and loaded body interfere. The resulting interferogram gives the loci of equal displacements of the body.

9.3.3 Double-Exposure Holographic Interferometry

In the double-exposure holographic interferometry, the hologram is produced by the reference light beam, the light beam of the unloaded body, and the light beam of the loaded body. In the reconstruction process, the hologram is illuminated by the reference beam. As a result, the images of the unloaded and loaded body are obtained. These two images interfere. The resulting interference pattern represents the loci of displacements of the body.

The double-exposure holographic interferometry compared with the real-time holographic interferometry has the disadvantage that only two forms of the body can be recorded in the hologram and compared in the reconstruction process, whereas in the real-time method the hologram constitutes the basis for comparison of one state of the body with a series of other states which may result, e.g., from application of a progressively increasing load. However, in the double-exposure method, the difficulties in real-time holographic interferometry associated with exposing the holographic plate at one time with the unloaded object and at another time with the loaded object are eliminated. The object is not used in the reconstruction process. Furthermore, the shrinkage of the emulsion of the holographic plate is the same for both waves. Therefore, the reconstructed waves are not distorted relative to one another.

9.3.4 Sensitivity Vector

In both, real-time and double-exposure holographic interferometry the waves emitted from the unloaded S_i and loaded S_n surfaces of the body are reconstructed and interfere. The resulting interference pattern represents the loci of the displacements of the body.

Let consider a light beam incident on the surface of the unloaded body along the direction of the unit vector n_i (**illumination direction**). Let the light beam that is scattered, reflected, or diffracted from the surface of the loaded body propagates along a direction with unit vector n_0 (**observation direction**) (Fig. 9.2). Let A and A'

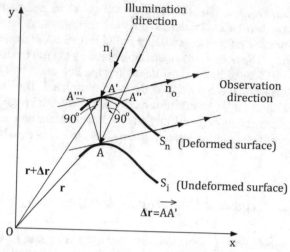

Fig. 9.2 Recording of the displacement of a surface from its undeformed S_i to its deformed S_n position. n_i and n_0 are unit vectors along the illumination and observation directions. $S = (n_i - n_0)$ is the sensitivity vector

be the position of a point on the surface S_i and S_n corresponding to the unloaded and loaded body. The difference s of the optical paths of the light rays along the direction of the unit vector n_0 (points A and A') is

$$s = AA'' - A'A''' = \Delta r \cdot n_i - \Delta r \cdot n_0 = \Delta r \cdot (n_i - n_0) = \Delta r \cdot S \qquad (9.17)$$

The vector $S = (n_i - n_0)$ is called the **sensitivity vector**. *It is the difference of the unit vectors along the illumination and observation directions.* The difference of the optical path of the two light rays along the observation direction of the unloaded and loaded body is the internal (dot) product of the sensitivity vector S and the displacement vector Δr of the body.

The two light rays along the observation direction of the unloaded and loaded body interfere. The interference pattern presents the loci of the optical-path difference s. The phase difference Δ between the two interfering rays is

$$\Delta = \frac{2\pi}{\lambda} s = \frac{2\pi}{\lambda} \Delta r \, (n_i - n_0) = \frac{2\pi}{\lambda} \Delta r \cdot S \qquad (9.18)$$

For the complete determination of the displacement vector of a point on the surface of the body three displacement components are needed. Therefore, Eq. (9.18) should be applied three times by changing the illumination and/or observation directions, for the complete determination of the three displacement components.

9.4 Holographic Photoelasticity

9.4.1 Introduction

Holographic photoelasticity is the application of holographic interferometry to photoelasticity. Using photoelastic materials interference patterns of superimposed families of isochromatics and isopachics are obtained. In the following, we will present the isochromatic-isopachic patterns obtained in real-time and double-exposure holographic photoelasticity.

9.4.2 Isochromatic-Isopachic Patterns

We will consider the real-time and double-exposure methods.

i. **Real-time holographic photoelasticity**

We will study the recording and reconstruction processes using the Jones calculus.
Recording process. Let a_1' and a_2' be the Jones vectors of the reference and object beams of the unloaded body. The intensity I of light transmitted by the hologram is

$$I = k(\tilde{a}_1' + \tilde{a}_2)(a_1' + a_2')$$

(9.19)

where k is a constant.

Reconstruction process. The hologram is illuminated by the reference beam and the beam of the loaded body with Jones vectors a_1' and a_3', respectively. The Jones vector a_T of the light that emerges from the hologram is

$$a_T = I(a_1' + a_3') = k(\tilde{a}_1' + \tilde{a}_2')(a_1' + a_2')(a_1' + a_3')$$

(9.20)

By omitting the terms in Eq. (9.20) which do not contribute to the interference pattern we obtain for the Jones vector

$$a_T' = k(\tilde{a}_1' a_1' + \tilde{a}_2' a_2')a_3' + k\tilde{a}_1' a_2' a_1'$$

(9.21)

In the above equation the quantities $\tilde{a}_1' a_1'$ and $\tilde{a}_2' a_2'$ represent the intensities of the reference and object beams for unloaded body. They can be taken as constants all over the field. Thus, Eq. (9.21) can be written as

$$a_T' = k'a_3' + k\tilde{a}_1' a_2' a_1'$$

(9.22)

where k' is a constant.

The light intensity I' that corresponds to vector a_T' is

$$I' = \tilde{a}'_T a'_T \tag{9.23}$$

Consider that the reference beam is circularly polarized. Its Jones vector is

$$a'_1 = \frac{1}{\sqrt{2}}\begin{bmatrix} -i \\ 1 \end{bmatrix} \tag{9.24}$$

Consider also that the object beam of the unloaded body is circularly polarized. Its Jones vector is

$$a'_2 = \frac{1}{\sqrt{2}}\begin{bmatrix} -i \\ 1 \end{bmatrix} \tag{9.25}$$

The Jones vector of the loaded body is

$$a'_3 = J\, a'_1 \tag{9.26}$$

where

$$J = \begin{bmatrix} e^{i\delta_1}\cos^2\theta + e^{i\delta_2}\sin^2\theta & \left(e^{i\delta_1} - e^{i\delta_2}\right)\cos\theta\sin\theta \\ \left(e^{i\delta_1} - e^{i\delta_2}\right)\cos\theta\sin\theta & e^{i\delta_1}\sin^2\theta + e^{i\delta_2}\cos^2\theta \end{bmatrix} \tag{9.27}$$

with

$$\delta_1 = \frac{2\pi}{\lambda}\Delta s_{t_1} = \frac{2\pi}{\lambda}(a_t\sigma_1 + b_t\sigma_1)d$$

$$\delta_2 = \frac{2\pi}{\lambda}\Delta s_{t_2} = \frac{2\pi}{\lambda}(b_t\sigma_1 + a_t\sigma_2)d \tag{9.28}$$

Introducing the values of a'_1, a'_2 and a'_3 from Eqs. (9.24), (9.25), and (9.26) into Eq. (9.23) we obtain the light intensity I' that contributes to the formation of the interference pattern

$$I' = C\,\cos\left[\frac{2\pi}{\lambda}\frac{a_t + b_t}{2}(\sigma_1 + \sigma_2)d\right]\cos\left[\frac{2\pi}{\lambda}\frac{a_t - b_t}{2}(\sigma_1 - \sigma_2)d\right] \tag{9.29}$$

Equation (9.28) indicates that *the obtained interference pattern consists of the superimposed families of isochromatics and isopachics.*

ii.   **Double-Exposure Holographic Photoelasticity**

Recording process. The hologram is formed by the reference beam a'_1, and the object beams of the unloaded a'_2 and loaded a'_3 body. The light intensity I transmitted through the hologram is

$$I = k_1\left(\tilde{a}'_1 + \tilde{a}'_2\right)\left(a'_1 + a'_2\right) + k'\left(\tilde{a}'_1 + \tilde{a}'_3\right)\left(a'_1 + a'_3\right) \tag{9.30}$$

Reconstruction process. The hologram is illuminated by the reference beam. The Jones vector a_T of the beam that emerges from the hologram is

$$a_T = I a_1' = \left[k_1 (\tilde{a}_1' + \tilde{a}_2)(a_1' + a_2') + k'(\tilde{a}_1' + \tilde{a}_3')(a_1' + a_3') \right] a_1' \qquad (9.31)$$

By omitting the terms in Eq. (9.31) which do not contribute to the interference pattern we obtain for the Jones vector a_T'

$$a_T' = k \tilde{a}_1' \, a_2' a_1' + k' \tilde{a}_1' a_3' a_1' \qquad (9.32)$$

The light intensity I' is

$$I' = \tilde{a}_T' a_T' \qquad (9.33)$$

Using the values of a_1', a_2', and a_3' from Eqs. (9.24), (9.25), and (9.26) we obtain for the light intensity I' (assuming $k = k'$)

$$I' == C \left\{ 1 + 2 \cos\left[\tfrac{2\pi}{\lambda} \tfrac{a_t + b_t}{2} (\sigma_1 + \sigma_2) d \right] \cos\left[\tfrac{2\pi}{\lambda} \tfrac{a_t - b_t}{2} (\sigma_1 - \sigma_2) d \right] + \atop \cos^2\left[\tfrac{2\pi}{\lambda} \tfrac{a_t - b_t}{2} (\sigma_1 - \sigma_2) d \right] \right\} \qquad (9.34)$$

where C is a constant.

Equation (9.34) indicates that *the obtained interference pattern consists of the superimposed families of isochromatics and isopachics.*

9.4.3 Generic Interpretation

We will provide a generic interpretation of the optical patterns obtained in holographic photoelasticity for both real-time and double-exposure methods.

In real-time holographic photoelasticity in the recording process, the hologram is formed by the reference beam and the object beam of the unloaded body. In the reconstruction process, the hologram is illuminated by the reference beam and the object beam of the loaded body. The illumination of the hologram by the reference beam gives as first-order diffraction the object beam of the unloaded body. The illumination of the hologram by the object beam of the loaded body gives as zero-order diffraction the object beam of the loaded body. These two beams, one of the body unloaded and the second of the body loaded interfere. The wave of the loaded body consists of two linearly polarized waves along the principal-stress directions. These two waves interfere with the wave of the unloaded body. Thus the irregularities of the unloaded body are removed from the interference pattern. The interference pattern consists of two superimposed families of curves that correspond to the loci of stress-optical retardations along the principal-stress directions. These two families of curves interfere according to the moiré effect, and the final pattern consists of the superimposed families of isochromatics and isopachics.

In double-exposure holographic photoelasticity in the recording process, the hologram is formed by the reference beam and the object beams of the unloaded and loaded body. In the reconstruction process, the hologram is illuminated by the reference beam. The object beams of the body loaded and unloaded are reconstructed and interfere. As before, the interference pattern consists of the superimposed families of isochromatics and isopachics.

Further Readings

1. DeVelis JB, Reynolds GO (1967) Theory and applications of holography. Addison Wesley, Boston
2. Stroke GW (1969) An introduction to coherent optics and holography, 2nd edn. Academic Press, New York
3. Kock WE (1969) Lasers and holography. An introduction to coherent optics. Doubleday & Co., New York
4. Smith HM (1969) Principles of holography. Wiley-Interscience, Hoboken
5. Robertson ER, Harvey JM (eds) (1970) The engineering uses of holography. Cambridge University Press, Cambridge
6. Caufield HJ, Lu, S (1970) The applications of holography. Wiley Interscience, Hoboken
7. Viénot JCM, Smigielski P, Royer H (1971) Holographic optique. Dunod, Paris
8. Collier RJ, Burckhardt, Lin LH (1971) Optical holography. Academic, New York
9. Kiemle H, Röss D (1972) Introduction to holographic techniques. Plenum Press, Cleveland
10. V. Brčić V (1975) Application of holography and hologram interferometry to photoelasticity. Springer, Berlin,
11. Theocaris PS, Gdoutos EE (1979) Matrix theory of photoelasticity. Springer, Berlin, pp 208–242
12. Vest JCM (1979) Holographic interferometry. Wiley, New York
13. Schumann W, Dubas M (1979) Holographic interferometry. Springer, Berlin
14. Yaroslavskii LP, Merzlyakov NS (1980) Methods of digital holography. Springer, Berlin
15. Kasper JE, Feller SA (1985) The complete book of holograms. Prentice Hall, Hoboken
16. Schumann W, Zürcher J-P, Cuche D (1985) Holography and deformation analysis. Springer, Berlin
17. Jones R, Wykes C (1989) Holographic and speckle interferometry, 2nd edn. Cambridge University Press, Cambridge
18. Ostrovskii L, Shchepinov VP, Yakovlev VV (1991) Holographic interferometry in experimental mechanics. Springer, Berlin
19. Ranson WF, Sutton MA, Peters WH (1993) Holographic and laser speckle interferometry. In Kobayashi AS (ed) Handbook of experimental mechanics 2nd ed. Society for Experimental Mechanics, pp 365–405
20. Rastogi PK (ed) (1994) Holographic interferometry. Springer, Berlin
21. Kreis T (1996) Holographic interferometry (Principles and methods) Akademie Verlag
22. Hariharan P (1996) Optical holography, 2nd edn. Cambridge University Press, Cambridge
23. Kreis T, Kreis T (1996) Holographic interferometry: principles and methods. Wiley-CVH
24. Tyler JR, Shelton JC (1997) Holographic interferometry. In: Shelton JC, Orr JF (eds) Optical measurement methods in biomechanics, Chapman & Hall, Boca Raton, pp 60–75
25. Cloud GL (1998) Optical methods in engineering analysis. Cambridge University Press, Cambridge, pp 343–391

26. Tonomura A (1999) Electron holography. Springer, Berlin
27. Khan AS, Wang X (2001) Strain measurements and stress analysis. Prentice Hall, Hoboken, pp 204–223
28. Sirohi RS, Osten W, Tay CJ, Shang HM, Chau FS (eds) (2001) Selected papers on holographic interferometry: applications. SPIE
29. Kreis T (2005) Handbook of holographic interferometry. Wiley-VCH
30. Ackermann GK, Eichler J (2007) Holography. Wiley-VCH
31. Barrekette ES, Bove VM (2008) Holographic imaging. Wiley Interscience
32. Pryputniewicz RJ (2008) Holography. In: Sharpe WN (ed) Handbook of experimental solid mechanics. Springer, Berlin, pp 675–700
33. Benton SA (2008) Holographic imaging. Wiley-Interscience. https://www.amazon.com/s/ref=dp_byline_sr_book_2?ie=UTF8&field-author=V.+Michael+Bove+Jr.&text=V.+Michael+Bove+Jr.&sort=relevancerank&search-alias=books
34. Schnars U, Jüptner W (2010) Digital holography: digital hologram recording, numerical reconstruction and related techniques. Springer, Berlin
35. Razumovsky IA (2011) Interference-optical methods of solid mechanics. Springer, Berlin, pp 37–82
36. Sciammarella CA, Sciammarella FM (2012) Experimental mechanics of solids. Wiley, New York, pp 631–679
37. Picart P, Li JC (2012) Digital holography. Wiley, New York
38. Toal V (2012) Introduction to holography. CRC Press, Boca Raton
39. Picart P, Li J-C (2013) Digital holography. Wiley, New York
40. Toker GR (2012) Holographic interferometry: A Mach–Zehnder approach. CRC Press, Boca Raton
41. Poon T-C, Liu J-P (2014) Introduction to modern digital holography. Cambridge University Press, Cambridge
42. Schnars U, Falldorf C, Watson J, Jüptner W (2014) Digital holography and wavefront sensing. Springer
43. Nehmetallah GT, Aylo R, Williams L (2015) Analog and digital holography with MATLAB. SPIE
44. Saxby G, Zacharovas S (2016) Practical holography, 4th edn. CRC Press, Boca Raton
45. Shimobaba T, Ito T (2019) Computer holography. CRC Press, Boca Raton
46. Matsushima K (2020) Introduction to computer holography. Springer, Berlin

Chapter 10
Optical Fiber Strain Sensors

10.1 Introduction

Fiber optic sensor (FOS) technology uses **optical fibers**. FOSs offers important advantages over conventional sensors, such as immunity to electromagnetic radiation, multiplexing, small size, high sensitivity, high accuracy, remote sensing, and are chemically and biologically inert. They have been employed for measurement of temperature, strain, index of refraction, humidity. Their development has been stimulated by the technological progress of fiber optic communication by providing higher performance, more reliable telecommunication links with decreasing bandwidth cost.

Optical fibers are made of glass or plastic with diameter of a human hair and they may be many miles long. Light is transmitted from one end of the fiber to the other and a signal may be imposed. Their operation is based on the phenomenon of total internal reflection. Fiber optic systems are superior to metallic conductors because it is possible to transmit a signal that contains more information than is possible with a metallic conductor.

In this chapter, we present the operation of optical fibers for transfer of light and describe the interferometric and Bragg grating fiber optic sensors for strain measurement.

10.2 Optical Fibers

10.2.1 Introduction

An **optical fiber** is a fundamental unit of any fiber-optics system. It is a strand of a dielectric material that can trap optical radiation at one end and guide it to the other. It is sometimes called *light pipe*. When a beam of light enters at one end of the fiber the light can be totally reflected internally if the angle of reflection is higher than a

© The Author(s), under exclusive license to Springer Nature Switzerland AG 2022
E. E. Gdoutos, *Experimental Mechanics*, Solid Mechanics and Its Applications 269,
https://doi.org/10.1007/978-3-030-89466-5_10

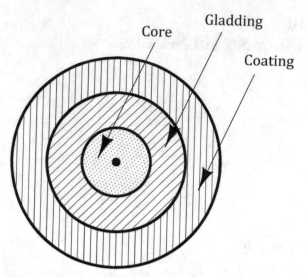

Fig. 10.1 An optical fiber consists of the core, the cladding, and an outer coating for protection

critical angle. Data are transmitted in optical fibers at a much higher rate and with less loss than an electrical signal in a copper wire. Optical fibers can support over one hundred wavelengths, each modulated to carry up to 10 gigabits (10^{10} bits) of information per second. For a hundred wavelengths that amount to a *terabit* (10^{12} bits) per second.

10.2.2 Structure

An optical fiber is cylindrical in shape. It consists at least of two optically dissimilar materials (Fig. 10.1). The inner material is called **core**. The other material that surrounds the core is called **cladding**. The energy is trapped in the core by reflection at the boundary of the core and the cladding. The cladding is surrounding by a coating of layers for strength and protection. The layers do not play any role in the guiding properties of the fibers. Usually, the fibers are arranged in bundles which consist of thousands of individual fibers. The diameter of the fibers is of the order of 0.002–0.01 mm.

10.2.3 Principle of Operation

The operation of optical fibers is based on the phenomenon of total internal reflection (Fig. 10.2). It occurs when a ray is incident on the interface between two materials

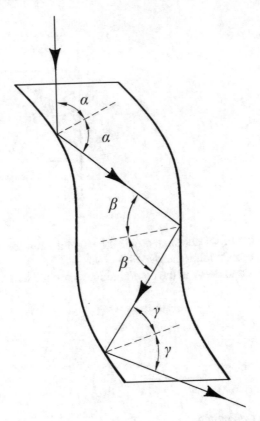

Fig. 10.2 Passage of a light ray through a transparent rod with index of refraction greater than the surrounding medium

with the second material having smaller index of refraction than that of the first material in which the ray is traveling. In optical fibers, the index of refraction of the core is higher than that of the cladding. Guidance is achieved by causing the electromagnetic beam to undergo a large number of reflections at the interface of the core and cladding materials. For optical fibers values of reflectivity for glass–glass boundaries of 0.9995 were established. Aluminum has a reflectivity of about 0.9.

Consider an optical fiber of cylindrical shape with circular cross section and let us study the propagation of a meridional ray through the core of the fiber (Fig. 10.3). The ray lies in the meridional plane which contains the cylindrical axis. The end faces of the fiber are normal to the fiber axis. A light ray enters the fiber from a medium of index of refraction n_0 and makes an angle θ with the fiber axis. The index of refraction of the core and gladding of the fiber is n_1 and n_2, respectively, with $n_1 > n_2$. By applying Snell's law at the end face of the fiber we obtain

$$n_0 \sin \theta = n_1 \sin \theta^1 = n_1 \cos \varphi \tag{10.1}$$

Fig. 10.3 Passage of a light ray through an optical fiber

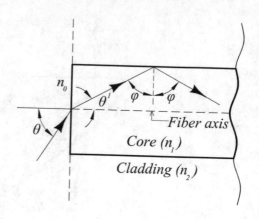

where θ^1 is the angle of refraction and φ is the angle the refracted ray makes with the normal to the core/gladding interface $(\varphi = 90° - \theta^1)$.

For total reflection to occur at the core/gladding interface we have

$$\sin \varphi_2 > n/n_1 \tag{10.2}$$

or

$$\cos \varphi < \left(1 - \frac{n_2^2}{n_1^2}\right)^{1/2} \tag{10.3}$$

Using Eq. (10.1) we obtain

$$\sin \theta < \frac{1}{n_0}\left(n_1^2 - n_2^2\right)^{1/2} \tag{10.4}$$

Equation (10.4) defines a critical angle of incidence θ_M given by

$$\sin \theta_M = \frac{1}{n_0}\left(n_1^2 - n_2^2\right)^{1/2} \quad \text{when } n_1^2 < n_2^2 + n_0^2 \tag{10.5}$$

or

$$\sin \theta_M = 1 \quad \text{when } n_1^2 > n_2^2 + n_0^2 \tag{10.6}$$

The above developments were made on the condition of total internal reflection at the core/cladding interface. Thus, Eq. (10.5) defines the **acceptable angle** $\theta = \theta_M$ (Fig. 10.4). It has the meaning: *for angles of incidence at the end faces of the fiber less than the acceptable angle, θ_M total reflection takes place at the core/cladding interface.*

Since the fiber is usually immersed in vacuum or air we can take $n_0 = 1$ and Eqs. (10.5) and (10.6) become

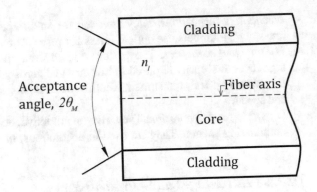

Fig. 10.4 Acceptance angle of an optical fiber

$$\sin \theta_M = \left(n_1^2 - n_2^2\right)^{1/2} \quad \text{when } n_1^2 < n_2^2 + 1 \tag{10.7}$$

or

$$\sin \theta_M = 1 \quad \text{when } n_1^2 > n_2^2 + 1 \tag{10.8}$$

Equations (10.7) and (10.8) give a measure of the light-gathering power of the fiber. The quantity given at the right-hand side of Eq. (10.7) is defined as the **numerical aperture(N.A.)** of the fiber. It is given by

$$\text{N.A.} = \left(n_1^2 - n_2^2\right)^{1/2} \tag{10.9}$$

The numerical aperture is a measure of the ability of an optical fiber to collect or confine the incident light ray inside it. It shows the efficiency with which light is collected inside the fiber in order to get propagated. It is among the most basic properties of optical fibers. *N.A.* is related to the acceptance angle θ_M which is the maximum angle through which light enters the fiber for total internal reflection.

10.2.4 Applications

Optical fibers transmit information in the form of light. They have revolutionized the world of network communication. Applications include:

- *Communication systems*: In various networking fields optical fibers increase the amount, speed, and accuracy of the transmission data. Internet and computer networking are common examples.
- *Medicine*: They are used in various instruments called *endoscopes* to view internal parts of the body. They can be inserted into the bronchial tubes, the colon, the bladder, and so on for visual examination. They are used to light the surgery area within the body, in microscopy and biomedical research.

- *Mechanical inspections*: They are used for on-site inspections for hard to reach areas, e.g., plumbers use optical fibers for inspection of pipes.
- *Military and aerospace*: They offer an ideal solution for data transmission with high level of security required in military and aerospace applications.
- *Broadcasting*: They transmit high definition television signals of higher bandwidth and speed.
- *Lightening and decoration*: They give an attractive, economical, and easy way of illuminating an area. They are used in decorations, like Christmas trees.

10.2.5   Advantages and Disadvantages

Advantages and disadvantages of optical fiber cables compared with metallic cables include:

Advantages

- *Greater bandwidth and faster speed*. Optical fibers support extremely high bandwidth and speed.
- *Smaller and light weight*. Optical fibers are thinner and can be drawn in smaller diameters than metallic cables. They better fit in cases where space is a concern.
- *Higher carrying capacity*. They carry a large amount of information in either digital or analog form.
- *Immunity to electromagnetic radiation*.
- *Interference*. Radar and other signals cannot interfere with the optical fibers.
- *Resistance to high temperatures*.
- *Transmittance of information over long distances*. They carry information without attenuation over greater distances than copper cables, which require boosters to be placed much closer.
- *Less signal degradation*. The loss of signal in optical fiber is less than that in copper cables.
- *Long lifespan*. They have a life cycle for over 100 years.

Disadvantages

- *Cost*. Copper cables are cheaper than optical fiber cables.
- *Prone to optical flux*. Optical fibers are prone to fiber flux at high optical powers, so they may get damaged.
- *Installation*. Installation of fiber cables is more labor intensive.
- *Fragility*. Optical fibers are rather fragile and of low resistance to twisting or bending.
- *Attenuation*. Loss of light in fibers due to scattering.
- *Vibrations*. Physical vibration will show up as signal noise.

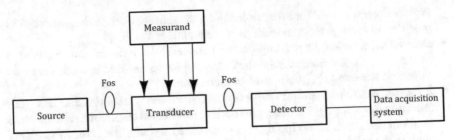

Fig. 10.5 Simplified architecture of an optical fiber sensor system

10.3 Fiber Optic Sensors (FOS)

10.3.1 Architecture of a FOS

A FOS is an element that translates changes of a measurand (strain in our case) into modulating (modifying) one or more parameters of light (intensity, wavelength, phase, polarization). The changes of the measurand are received by a detector that measures these changes (Fig. 10.5). The building block of a FOS system is analogous to that of an electrical resistance strain gage (ERSG). It consists of the following four components:

- *Light source.* A laser or broadband light source is used. It is analogous to the voltage used in ERSGs.
- *Communication channel*: This is the optical fiber. In some cases, the fiber can serve as both the channel and the sensor. It is analogous to the Wheatstone bridge and the electrical wires connecting to the ERSG.
- *Optical transducer*: This is the optical fiber sensor. It is analogous to the ERSG.
- *Data acquisition system*: Its function is to detect the changes induced by the transducer in the input light. It is a duplicate of the acquisition system of ERSGs.

10.3.2 Classification of FOSs

FOSs can be classified into the following categories regarding the criterion chosen. We use four criteria: the induction nature, the spatial distribution, the induced modulation, and the reflection or transmission.

- *Induction nature*: FOSs may serve as both conduits for transferring the signal and sensors. In this case, they are called **intrinsic**. When the optical fiber delivers the signal to the sensor the FOS is called **extrinsic**.
- *Spatial distribution*: FOSs maybe point, distributed, and cumulative. A **point** FOS provides measurement of the measurand at one location. A **distributed** FOS

provides measurement for every point along the sensing FOS. A cumulative FOS provides an average measurement over the length of the FOS.

- *Induced modulation*: Depending on the parameters of light that are modulated by the changes of the measurand the FOSs can be categorized to those based on **intensity, wavelength, phase, and polarization.**
- *Reflection or transmission*: FOSs work in **reflection** or **transmission** mode. In reflection mode the light traveling in the fiber is reflected back to the source, while in the transmission mode the light is transmitted through the fiber. Most FOSs work in the reflection mode.

10.3.3 Interferometric FOSs

As we have discussed in Chapter 8, interferometry has become a standard method for measurement of small distance changes. Interferometric systems are generally restricted to laboratories as alignment of the optical components and beams can be easily disturbed by random vibrations. Fiber optic versions of classical interferometers allow the construction of interferometric systems for in-situ measurement of quantities such as strain, temperature, etc. outside of the laboratory. *Fiber optic interferometers are analogs of classical interferometers.* They use the interference between two beams that propagate through different optical paths of a single fiber or two different fibers. The interferometers divide the light from a source into two different beams which follow different paths and they recombine. In the following, we will present the Mach–Zehnder, Michelson, and the Fabry-Perrot fiber-optics interferometers and the polarization sensors.

Mach–Zehnder Interferometer (MZI). It has two different arms, the reference arm and the sensing arm (Fig. 10.6). The incident light is split into two arms by a fiber coupler. The two beams recombine by another fiber coupler. The recombined beams have a difference in the optical path length (OPL). The reference arm does not have any variation of OPL. The sensing arm is exposed to the variation of OPL induced by strain, temperature, index of refraction, etc. Due to the difference in the OPL between the two beams of the interferometer, an interference pattern is formed which is detected and analyzed.

Michelson Interferometer (MI). It is quite similar to MZI. Light is split into two arms and each beam is reflected at the end of each arm (Fig. 10.7). The two mirrors

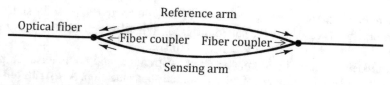

Fig. 10.6 Optical fiber Mach–Zehnder interferometer

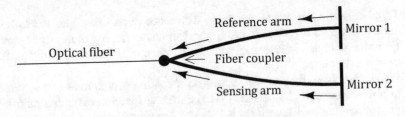

Fig. 10.7 Optical fiber Michelson interferometer

at the ends of the two arms are the main difference between MZI and MI. MI is like a half of MZI. After traversing the fibers the light is reflected from the mirrors at the ends of the fibers and propagates back to a coupler where interference occurs.

Fabry–Perot Interferometer (PFI). It is composed of a cavity with two parallel reflecting surfaces (Fig. 10.8). Interference fringes are formed when light is reflected from the two surfaces of the cavity. They are related to the change of the length of the cavity and the index of refraction. It comes in two forms, the extrinsic and the intrinsic.

In the extrinsic sensor, the cavity is formed out of the fiber by a supporting structure as is shown in Fig. 10.8a. In the intrinsic fiber, the sensor has two mirrors formed within the fiber (Fig. 10.8b).

The phase difference of the FPI sensor is given by

$$\Delta = \frac{2\pi}{\lambda} n(2L) \tag{10.10}$$

where n is the index of refraction of the material of the cavity and L is the length of the cavity.

For an extrinsic sensor with air cavity $n = 1$. For a strain sensor the length of the cavity L and the index of refraction n (in the case of the intrinsic sensor) change due

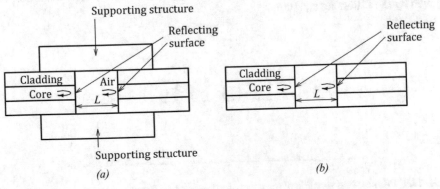

Fig. 10.8 Optical fiberFabry-Perror interferometer. Extrinsic (**a**), Intrinsic (**b**)

to the applied strain. Therefore, the phase difference Δ changes. The new index of refraction under load is related to the applied strain, according to the strain-optical law. From the obtained interference pattern the change of length L, and therefore, the strain is determined.

Polarization sensors. They are based on the same principle as stress induced optical birefringence of photoelastic materials. They function by taking light from a single-mode fiber and launching it into the length of the sensor. Light travels along the fast and slow birefringent axes of the sensor. When a load has applied the birefringence between the two axes changes in accordance with the photoelastic law. The output of the sensor is launched back into a single-mode fiber generating an interference signal. The optical arrangement of the polarization sensor is analogous to the plane polariscope in photoelasticity.

10.3.4 Fiber Bragg Grating Sensors (FBGSs)

A fiber **Bragg grating sensor** is an optical device that presents a periodic variation of the refractive index of the core along the grating length. For an equally spaced modulation the index of refraction n_c varies along the grating length as

$$n_c = n_{c0} \text{ for } 0 < z < \Lambda/2$$
$$n_\varepsilon = n_{c0} + \Delta n \quad \text{for } \Lambda/2 < z < \Lambda \tag{10.11}$$

where n_{c0} is the index of refraction of the unmodulated fiber, Δn is the change of the index of refraction and Λ is the grating pitch (Fig. 10.9).

When light travels inside the grating some portion of the light is reflected back from each grating plane. When a special (Bragg) condition is satisfied the reflected portions of light from each grating plane add up to form a backward reflected beam with wavelength λ_B. The grating works as a mirror that reflects a selected (Bragg) wavelength λ_B and transmits the remaining (Fig. 10.10). The Bragg condition is given by the following equation

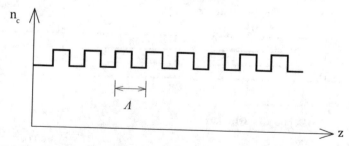

Fig. 10.9 Profile of the variation of the index of refraction of a Bragg grating sensor

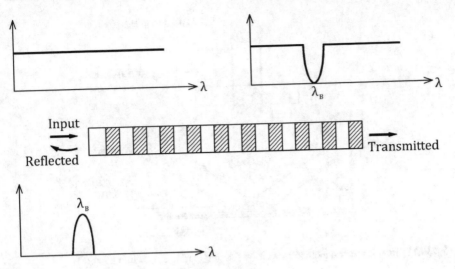

Fig. 10.10 Operating principle of a Bragg grating sensor

$$\lambda_B = 2n_{\text{eff}}\Lambda \tag{10.12}$$

where n_{eff} is the effective index of refraction of the fiber core.

When the Bragg condition is not satisfied the reflected light from the grating planes cancel out and no reflection is observed.

By differentiating Eq. (10.12) with respect to the length l of the FBGS we obtain for the shift of the Bragg wavelength due to an applied strain

$$\Delta\lambda_B = 2\left(\Lambda\frac{\partial n_{\text{eff}}}{\partial l} + n_{\text{eff}}\frac{\partial \Lambda}{\partial l}\right) \tag{10.13}$$

where $\partial n_{\text{eff}}/\partial l$ is the variation of the effective index of refraction n_{eff} and $\partial \Lambda/\partial l$ is the change of the pitch Λ of the fiber with respect to the fiber length l.

Introducing the strain-optical relations of the fiber Eq. (10.13) can be written as

$$\frac{\Delta\lambda_B}{\lambda_B} = (1 - p_e)\varepsilon \tag{10.14}$$

where p_e is a strain-optical coefficient and ε is the strain.

For a germanium doped silica fiber $p_e = 0.22$, and Eq. (10.14) becomes

$$\frac{\Delta\lambda_B}{\lambda_B} = 0.78\varepsilon \tag{10.15}$$

Equation (10.15) is used for the determination of strain ε from the shift $\Delta\lambda_B$ of the Bragg wavelength λ_B. At a wavelength range of 1550 nm, we obtain from Eq. (10.15)

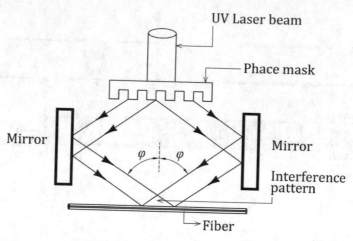

Fig. 10.11 Interferometric method to inscribe a Bragg grating sensor into a fiber

$$\frac{\Delta\lambda_B}{\varepsilon} = 1.2 \; pm/\mu\varepsilon \quad \left(1 \; pm = 10^{-12} \; m, \; 1\mu = 10^{-6}\right) \tag{10.16}$$

By a similar approach, we can use FBGSs to measure other physical quantities, like temperature, pressure, vibration, load.

Fabrication (called *inscription* or *writing*) of FBGS is simple. It is made by a systematic change of the index of refraction of the core of the optical fiber. This is made possible by a phenomenon known as **photosensitivity**. The fiber is exposed to intense ultra-violet (UV) radiation. Two main methods are used: interference and masking. In the interference method, an interference pattern is produced and it is focused on the region of the core of the fiber. The index of refraction changes at areas of highest intensity. The amount of change depends on the intensity and duration of the exposure as well as the photosensitivity of the fiber (Fig. 10.11). In the masking method, a proper mask that defines the period of the FBGS is used.

10.3.5 *Multiplexing*

Multiplexing is the process by which two or more signals are transmitted through a single communication channel. It offers the ability for multiple parameters to be monitored not only by a single instrument but also with all the data transmitted on a single piece of an optical fiber. Multiplexing is a great advantage of FBGSs. Many sensors can be multiplexed to provide measurements of strain in a structure for health monitoring. There are three main arrangements for multiplexing of FBGSs: wavelength division (WDM), time division (TDM), and frequency division (FDM).

In **wave division multiplexing (WDM)** each FBGS is written at a slightly different wavelength λ_B. The reflective spectrum of the multiplexed sensors contains a series of peaks according to the Bragg Eq. (10.12). A broadband light source is used to interrogate an array of FBGSs arranges in series. The major drawback of this type of multiplexing is the overlapping of the Bragg wavelengths λ_B of the various sensors.

In **time division multiplexing (TDM)** each FBGS is identified by the time required for light to travel from the source to the detector across the fiber. If the sensors are separated by an adequate length of optical fiber, usually 1 m, this time is enough to separate the responses of the array of sensors.

In **frequency division multiplexing (FDM)** each sensor is identified by the particular modulation frequency.

10.3.6   Advantages and Disadvantages of FOSs

Advantages

Fiber-optics sensors (FOSs) and particularly fiber Bragg grating sensors have big advantages over electrical equivalents. They are the only choice of election if the number of sensors is high (let say more than 30 sensors) or if the distances between the sensors are long (of kilometers). Other advantages of FOSs include:

- Immunity to electromagnetic interference and radio frequency interference.
- Insensitiveness to lightning strikes and explosive atmospheres.
- Safety to be used in harsh environments (high temperature above 1400 °C).
- Large bandwidth.
- Small size and weight.
- High multiplexing capability allowing installation of a large number of FOSs to a single optical fiber.
- Remote sensing capability (large distance between sensor and interrogator of kilometers).
- High resistance to fatigue and mechanical failure.
- Measurement of many physical quantities, including strain, temperature, displacement, flow, rotation, velocity, acceleration, radiation, humidity, pH, etc.

Disadvantages

Disadvantages include:

- They are expensive.
- Detection systems are complex.
- Require precise installation methods.
- Require training of the user.

10.3.7   Applications of FOSs

FOSs are used in a variety of applications such as:

- Health monitoring of structures (buildings, bridges, tunnels, dams).
- Marine applications (ships, submarines, etc.).
- Oil and gas applications.
- Wind power applications.
- Composite materials and structures (they can easily be embedded in the manufacturing of structural parts).

Further Readings

1. Chester AN, Martellucci S, Verga Scheggi AM (eds) (1987) Optical fiber sensors. Springer, Berlin
2. Arditty HJ, Dakin JP, Kersten RT (eds) (1989) Optical fiber sensors. Springer, Berlin
3. Grattan LS, Meggitt BT (eds) (1995) Optical fiber sensor technology. Springer Science+Business Media, Berlin
4. Grettan KTV, Meggitt BT (2000) Optical fiber sensor technology. Springer Science and Business Media, Berlin
5. Glisic B, Inaudi D (2007) Fiber optic methods for structural health monitoring. Wiley, New York
6. Baldwin CS (2008) Optical fiber strain gages. In: Sharpe WN (ed) Handbook of experimental solid mechanics. Springer, Berlin, pp 347–370
7. Yin S, Ruffin YP, Yu FTS (2008) Fiber optic sensors, 2nd edn. CRC Press, Boca Raton
8. Kashyap R (2009) Fiber Bragg gratings, 2nd edn. Academic, New York
9. Udd E, Spillman WB (2011) Fiber optic sensors. Wiley, New York
10. Fang Z, Chin K, Qu R, Cai H (2012) Fundamentals of optical fiber sensors. Wiley, New York
11. Kang JU (ed) (2013) Fiber optic sensing and imaging. Springer, Berlin
12. Krohn DA, MacDougall TW, Mendez A (2014) Fiber optic sensors. SPIE
13. Rajan G (ed) (2015) Optical fiber sensors. CRC Press, Boca Raton
14. Santos JL, Farahi F (eds) (2015) Handbook of optical sensors. CRC Press, Boca Raton
15. Hartog AH (2017) An introduction to distributed optical fibre sensors. CRC Press, Boca Raton
16. Rao Y-J, Ran Z-L, Gong Y (2017) Fiber-optic Fabry-Perrot sensors. CRC Press, Boca Raton
17. Rajan G, Prusty BG (2017) Structural health monitoring of composite structures using fiber optic methods. CRC Press, Boca Raton
18. Matias IR, Ikezawa S, Corres J (eds) (2017) Fiber optic sensors. Springer, Berlin
19. Werneck MM, Allil R, Nazaré F (2017) Fiber Bragg gratings. Theory, fabrication and applications. SPIE Press
20. Oliveira R, Botas Bilro LM, Nogueira RN (2019) Polymer optical fiber Bragg gratings. CRC Press, Boca Raton
21. Novais S, Ferreira M., Pinto J (2019) Optical fiber sensors for challenging media. PAL Lambert Academic Publishing CRC Press, Boca Raton
22. Du Y, Sun B, Li J, Zhang W (2019) Optical sensing and structural health monitoring technology. Springer, Berlin
23. Kulkarni V (2020) Propagation effects in optical fibers and fiber optic sensor systems. PAL Lambert Academic Publishing
24. Werneck MM, Célia SB, Allil, R (eds) (2020) Plastic optical fiber sensors. CRC Press, Boca Raton
25. Hisham HK (2020) Fiber Bragg grating sensors. CRC Press, Boca Raton

26. Wei L (ed) (2020) Advanced fiber sensing technologies. Springer, Berlin
27. Del Villar I, Matias IR (eds) (2021) Optical fibre sensors. Wiley, New York
28. Hill KO, Meltz G (1997) Fiber Bragg grating technology fundamentals and overview. J Lightwave Tech 15:1263–1276
29. Askins CG, Putnam MA, Friebele EJ (1997) Fiber grating sensors. J Lightweight Tech 15:1442–1463
30. Tennyson RC, Coroy T, Duck G, Manuelpillai G, Mulvihill P, Cooper DJF, Smith PWE, Mufti AA, Jalali SJ (2000) Fibre optic sensors in civil engineering structures. Can J Civ Eng 27:880–889
31. Grattan KTV, Sun DT (2000) Fiber optic sensor technology; an overview. Sens Actuators 82:40–61
32. Lee B (2003) Review of present status of optical fiber sensors. Opt Fiber Tech 9:57–79
33. Lee BH, Kim YH, Park KS, Eom JB, Kim MJ, Rho BS, Choi HY (2012) Interferometric fiber optic sensors. Sensors 12:2467–2486
34. Li K (2015) Review of the strain modulation methods used in fiber Bragg grating sensor. J Sensors: Article ID 1284520. https://doi.org/10.1155/2016/1284520
35. Campanella CE, Cuccovillo A, Campanella C, Yurt A, Passaro MN (2018) Fibre Bragg grating based strain sensors: review of technology and applications. https://doi.org/10.3390/s18093115
36. Sahota JK, Gupta N, Dhawan D (2020) Fiber Bragg grating sensors for monitoring of physical parameters: a comprehensive review. Opt Eng 59(060901):1–35
37. Dong L, Gang T, Bian C, Tonga R, Wang J, Hu M (2020) A high sensitivity optical fiber strain sensor based on hollow core tapering. Opt Fiber Tech 56. https://doi.org/10.1016/j.yofte.2020.102179

Chapter 11
Speckle Methods

11.1 Introduction

Speckle methods are high-sensitivity non-contact optical methods for measuring displacements. They are based on the speckle effect. Speckles are granular dots that result from the illumination of a diffusively reflecting rough surface with coherent light. The reflected or scattered wavelets interfere to create a random speckle pattern with statistical properties. The laser speckle is due to the coherence of the light and the roughness of the surface of the order of the wavelength of light.

Even though the earliest observations about the speckle effect were made by Newton towards the end of the seventeenth century the speckle effect first came to prominence with the invention of the laser. It was first regarded purely as a nuisance. Images of objects illuminated by laser were covered with a grainy structure. It was later realized that the speckles can be studied for their own sake and became the basis of a series of optical methods in experimental mechanics grouped under the name "speckle methods".

Speckles can be used as "fingerprints" to impose a texture on a surface. For displacement measurement, two exposures, one for the specimen undeformed and a second for the specimen deformed are superimposed. The method of **speckle photography** is based on this principle. The sensitivity and range of speckle photography can be increased by adding a reference beam resulting in the method known as **speckle interferometry**. With the implementation of video and digital techniques a method known as **electronic speckle pattern interferometry(ESPI)** was developed. For the measurement of first derivatives of surface deformations a full-field, non-contact high-sensitivity interferometric method named **shearography** was developed. The interference is generated by two identical laterally sheared object beams, and no reference beam is required.

In the following, we present the speckle effect and the methods of speckle photography, speckle interferometry, shearography, and electronic speckle pattern interferometry (ESPI) for the measurement of displacements.

© The Author(s), under exclusive license to Springer Nature Switzerland AG 2022
E. E. Gdoutos, *Experimental Mechanics*, Solid Mechanics and Its Applications 269,
https://doi.org/10.1007/978-3-030-89466-5_11

11.2   The Speckle Effect

A speckle pattern is produced when coherent light is reflected or scattered from different parts of a surface. Consider a surface that is rough on the scale of the wavelength of light illuminated by coherent light. Rays from different parts of the surface travel different optical path lengths when impinging on an image plane (Fig. 11.1). The waves have the same frequency, but different phases and amplitudes. A point on the image plane is illuminated by many waves arriving from different points of the illuminated area. The resulting intensity at the point is determined by the coherent addition of the waves associated with the light rays reflected or scattered by various points of the surface. The light intensity on the image plane varies from one point to another and it is the cause of formation of laser speckles. This type of speckle is called **objective speckle**.

The statistical average diameter S_{obj} of the objective speckle is given by

$$S_{obj} = 1.22 \frac{L}{D} \lambda \tag{11.1}$$

where L is the distance between the object and the image plane, D is the diameter of the illuminated area of the object and λ is the wavelength of light (Fig. 11.2).

Consider that the system incorporates a lens (Fig. 11.3). The pattern observed on the image plane is called **subjective speckle** pattern because the structure of the speckle depends on the parameters of the viewing system. For instance, the size of the speckle depends on the size of the lens aperture.

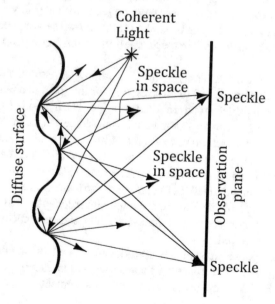

Fig. 11.1 Formation of speckles (objective) on an image plane by interference of light rays scattered or reflected form a surface

Object

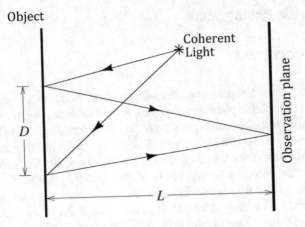

Fig. 11.2 Formation of an objective speckle and related parameters

Object

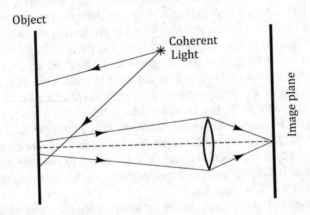

Fig. 11.3 Formation of a subjective speckle

The statistical average diameter S_{subj} of the subjective speckle is given by

$$S_{subj} = 1.22(1 + M)\lambda \frac{F}{M} \tag{11.2}$$

where M is the magnification of the lens and $F = f/D$ where f is the focal length and D is the aperture diameter of the lens.

In the following, *we will deal with subjective speckles only.*

11.3   Speckle Photography

11.3.1   Introduction

The speckle pattern is a signature on the surface of the body under study. When the surface moves the speckle pattern also moves. A double-exposure photograph called **specklegram** of the surface corresponding to the undeformed and the deformed states of the body produces two speckle patterns. Knowledge of the location of the speckles in each speckle pair gives the displacement of points of the surface. No fringes will result directly from the superposition of two speckle patterns. Optical Fourier processing and spatial filtering techniques are needed to reveal the hidden information. The obtained fringes can be interpreted as contours of displacements. Two optical techniques have been developed to measure the separation of the speckles. A point-by-point technique that produces Young's fringes with spacing and direction dependent on the displacement of the points of the surface between the two exposures and a spatial filtering technique to obtain whole-field displacement fringes.

Even though coherent light is necessary for the production of the speckle pattern the actual mechanism of speckle photography is incoherent. Only the intensity, not the phase distribution of the two speckle patterns, is recorded on the specklegram. The two speckle patterns on the specklegram need to remain correlated (in touch). The displacements of the speckles must be sufficient to produce a separated pair of speckles. On the other hand, the displacements should not be too much far apart for the two speckle patterns to get decorrelated. This limits the sensitivity of the method to approximately one speckle diameter. The method of speckle photography is insensitive to out-of-plane displacements. It is limited to in-plane displacements only. The method is less sensitive to vibrations compared to other coherent methods, like interferometry and holography.

The specklegram can be considered as a grating whose pitch and orientation vary from point to point. In this respect, there is an analogy between moiré and speckle methods. The moiré methods can be considered as special cases of speckle methods.

11.3.2   Point-by-Point Interrogation of the Specklegram

A double-exposure specklegram is a complex diffraction grating of varying pitch and orientation. It consists of many speckle pairs corresponding to the undeformed and deformed states of the body. The spacing and orientation of two corresponding speckles indicate the magnitude and orientation of the local displacement. Measurement of displacements between local speckle pairs cannot be made practically. To determine the speckle pair separations the diffraction characteristics of the specklegram are used. The Fourier transform of the specklegram contains Young's interference fringes which, as it was developed in Sect. 2.9.4, can be obtained by illuminating the specklegram with a laser beam and observing the far field diffraction pattern.

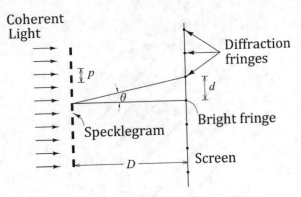

Fig. 11.4 Formation of Young's fringes by illuminating a speckle grating by a coherent beam and related parameters

For a small piece of the specklegram, the speckle pairs form a grating of parallel lines with pitch p. When this piece of specklegram is illuminated by a laser point light source the diffraction pattern consists of Young's parallel fringes. According to the grating Eq. (2.50) for a screen placed at a distance D from the grating we have (Fig. 11.4)

$$\sin \theta = \tan \theta = \frac{d}{D} = \frac{\lambda}{p} \tag{11.3}$$

where θ is the angle made by the normal to the grating and the first fringe, d is the spacing of the dark fringes and λ is the wavelength of light.

From Eq. (11.3) we obtain for the grating pitch p which is the displacement between two corresponding speckles at the specklegram

$$p = \frac{\lambda D}{d} \tag{11.4}$$

The obtained diffraction Young's fringes are perpendicular to the displacement vector (Fig. 11.5). Thus, the point-by-point interrogation of the specklegram with a point laser beam provides the magnitude and direction of the displacement.

11.3.3 Spatial Filtering of the Specklegram

The above point-by-point interrogation of the specklegram is tedious and does not provide whole-field displacement fringes. The double-exposure specklegram is a complicated diffraction grating. It is composed of many small pieces of gratings of variable pitch and orientation. It contains a wealth of information for measuring

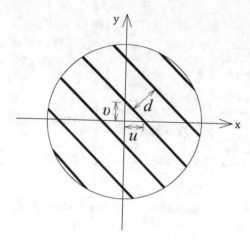

Fig. 11.5 Young's fringes corresponding to speckle pairs

displacements. However, it needs further processing to extract this information. This can be achieved by using the spatial filtering technique developed in Sect. 2.9.7.

When the entire specklegram is illuminated by a coherent light beam the irradiance distribution on the image plane will consist of the Fourier transforms of all individual small diffraction gratings of different pitch and orientation. It will be a complicated no recognizable structure of no use.

Consider all points on the specklegram that have the same displacement vector. When these pairs of speckles are illuminated by a pen coherent light the result on the image plane is a set of parallel fringes. Their spacing is related to the distance p (displacement) between the speckle pairs by Eq. (11.4). Their orientation is perpendicular to the direction of the displacement. To perform the Fourier transfer of the diffraction gratings in laboratory dimensions let us image the diffraction pattern by a lens of local length F. If l is the distance on the image plane of Young's fringes from an origin, then $l = nd$, where n is the fringe order and d is the distance between fringes. Then Eq. (11.4) can be written as

$$p = \frac{nF\lambda}{l} \tag{11.5}$$

Equation (11.5) is used for the determination of the distance of separation of the speckle pairs which is the in-plane displacement.

The above discussion suggests that if the specklegram is filtered by placing a spatial filter aperture on the transform plane at a distance l from the origin (Fig. 11.6) the resulting fringe pattern will present the fringes of equal displacement components given by Eq. (11.5). The direction of the displacements is along the radial distance from the origin to the aperture. Equation (11.5) indicates that the magnitude of the displacement is inversely proportional to the distance l of the aperture from the origin.

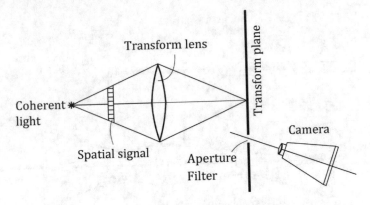

Fig. 11.6 Filtering a specklegram by placing a spatial filter on the transform plane

Thus, *to obtain the two in-plane displacement components along the x- and y-axis we place on the image plane a spatial aperture filter along the x- and y-axis respectively at different distances from the origin. The obtained diffraction fringes are the loci of equal displacement components along the x- and y-axis. The displacements are determined from Eq.* (11.5).

The above analysis is based on physical arguments. The same result is obtained by mathematical analysis of the intensity distribution on the image plane [1]. The displacements u_x and u_y along the x- and y-axis are given by

$$u_x = \frac{nF\lambda}{l_x}, \quad u_y = \frac{nF\lambda}{l_y} \tag{11.6}$$

where l_x and l_y are the distances of the aperture from the origin along the x- and y-axis, respectively.

Displacement components along any direction can be obtained by placing a spatial filter aperture along that particular direction. Note that Eq. (11.6) for the determination of displacements u_x and u_y is identical in form to the moiré fringe Eq. (3.1) with equivalent grating pitches being $p_1 = F\lambda/r_x$ and $p_2 = F\lambda/r_y$.

11.4 Speckle Interferometry

In speckle interferometry, a coherent reference beam or a second speckle pattern is coherently mixed with the object speckle pattern. The sensitivity of the technique is of the order of the wavelength of light, as in classical interferometry. The displacement difference between two successive fringes is equal to one wavelength of the light used. The use of a reference beam preserves the phase information in the speckle pattern.

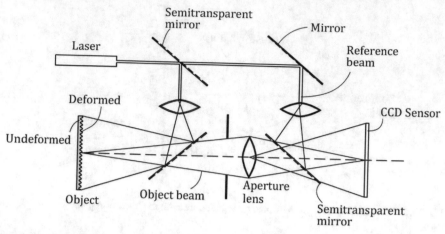

Fig. 11.7 Experimental arrangement of speckle interferometry

An experimental arrangement of the method is shown in Fig. 11.7. A reference beam is superimposed on the object beam to create an interfered speckle pattern on the image plane. Let the object beam for the undeformed body and the reference beam are represented by the waves $ae^{i\theta}$ and $a_r e^{i\theta_r}$, respectively. Then, the light intensity I at a point $P(x, y)$ of the specklegram is given by

$$I = \left(ae^{i\theta} + a_r e^{i\theta_r}\right)\left(ae^{i\theta} + a_r e^{i\theta_r}\right)^* \tag{11.7}$$

where * represents the conjugate of the respective quantity.

Equation (11.7) becomes

$$I = a^2 + a_r^2 + 2aa_r \cos(\theta - \theta_r) = a^2 + a_r^2 + 2aa_r \cos\varphi_r$$

where $\varphi_r = \theta - \theta_r$.

Let the object beam of the deformed body is represented by the wave $a'e^{i\theta'}$. Then, the intensity I' at the same point $P(x, y)$ is

$$I' = a'^2 + a_r^2 + 2aa_r \cos\varphi'_r = a'^2 + a_r^2 + 2a'a_r \cos(\Delta + \varphi_r) \tag{11.8}$$

where $\varphi'_r = \theta' - \theta_r$, $\Delta = \theta' - \theta$ is the phase change between the deformed and the undeformed states of the body.

Two speckle patterns are obtained, one with the body undeformed and a second with the body deformed. The intensity I_s at point $P(x, y)$ is the difference of the intensities for the two states, undeformed and deformed, of the body. We have from Eqs. (11.7) and (11.8) by neglecting the change of the amplitude of the object waves between the two states $(a' \approx a)$

$$I_s = |I' - I| = 2aa_r|\cos\varphi_r - \cos(\Delta + \varphi_r)| = 4aa_r\left|\sin\left(\varphi_r + \frac{\Delta}{2}\right)\sin\frac{\Delta}{2}\right|$$

(11.9)

From Eq. (11.9) it is shown that the change of intensity of the double-exposure specklegram depends on the factor $\sin(\Delta/2)$. Black speckles will be obtained for I_s = 0, that is, $\Delta = 2\pi k$ ($k = 0, 1, 2, \ldots$). These speckles belong to the same fringe.

Two speckle patterns are obtained, one before and a second after the deformation of the body. In the reconstruction process, the developed specklegram is illuminated by the same reference beam to obtain the fringes.

The phase difference Δ is related to the sensitivity vector $S = (n_i - n_0)$ by (see Eq. 9.18)

$$\Delta = \frac{2\pi}{\lambda}s = \frac{2\pi}{\lambda}\Delta r.S = \frac{2\pi}{\lambda}\Delta r.(n_i - n_0)$$

(11.10)

where Δr is the displacement vector and n_i *and* n_0 are the unit vectors along the illumination and observation directions, respectively.

The sensitivity vector lies along the bisector of the angle between the illumination and observation directions. It can be assumed that it is constant for all points of the surface of the body under study due to the large distances between the illumination source, the body, and the image plane.

Consider now a plate on the xy plane illuminated by a ray at an angle θ with the z-axis and a speckle image that is formed by a lens on the image plane (Fig. 11.8). Let the plate moves along the z-axis perpendicular to the plane of the plate by Δz. The observation direction is along the z-axis. We have for the unit vectors n_i *and* n_0

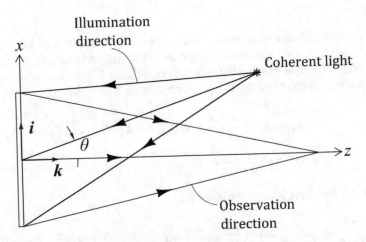

Fig. 11.8 A plate illuminated by a ray at an angle θ with the z-axis. The observation direction is along the z-axis

$$n_0 = k, n_i = -\cos\theta k - \sin\theta i \tag{11.11}$$

where i and k are the unit vectors along the x- and $z$axis, respectively.
The sensitivity vector S is

$$S = n_i - n_0 = -(1 + \cos\theta)k - \sin\theta i \tag{11.12}$$

The displacement vector Δr is

$$\Delta r = \Delta z\, k \tag{11.13}$$

The phase difference Δ is obtained from Eq. (11.10) as

$$\Delta = \left|\frac{2\pi}{\lambda}\Delta r.(n_i - n_0)\right| = \frac{2\pi}{\lambda}\Delta z k.[(1 + \cos\theta)k + \sin\theta i]$$
$$= \frac{2\pi}{\lambda}(1 + \cos\theta)\Delta z \tag{11.14}$$

Dark fringes are obtained for $\Delta = 2n\pi$ ($n = 1, 2,...$). Then, Eq. (11.14) renders

$$\Delta z = \frac{n\lambda}{1 + \cos\theta} \tag{11.15}$$

Equation (11.15) gives the contours of displacement Δz. When the light impinges on the specimen perpendicularly ($\theta = 0$) Eq. (11.15) becomes

$$\Delta z = \frac{n\lambda}{2} \tag{11.16}$$

For the determination of in-plane displacements, we consider two object beams 1 and 2 illuminating the specimen at equal angles θ with the z-axis in the xz-plane. (Fig. 11.9). No reference beam illuminates the specimen. Let $\Delta r = d = d_x i + d_y j + d_z k$ be the displacement vector. The phase change for beam 1 is

$$\Delta_1 = \frac{2\pi}{\lambda}\Delta r.(n_i - n_0) = \frac{2\pi}{\lambda}(d_x i + d_y j + d_z k).[(1 + \cos\theta)k + \sin\theta i]$$
$$= \frac{2\pi}{\lambda}[d_x \sin\theta + d_z(1 + \cos\theta)] \tag{11.17}$$

Similarly, the phase change for beam 2 is

$$\Delta_2 = \frac{2\pi}{\lambda}\Delta r.(n_i - n_0) = \frac{2\pi}{\lambda}(d_x i + d_y j + d_z k).[(1 + \cos\theta)k - \sin\theta i]$$
$$= \frac{2\pi}{\lambda}[-d_x \sin\theta + d_z(1 + \cos\theta)] \tag{11.18}$$

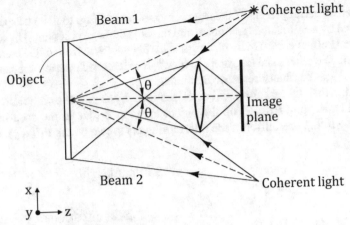

Fig. 11.9 Determination of in-plane displacements by illuminating the specimen with two object beams 1 and 2 at equal angles θ with the z-axis

The phase change between the two beams is

$$\Delta = \Delta_1 - \Delta_2 = \frac{4\pi}{\lambda} d_x \sin\theta \qquad (11.19)$$

Dark fringes are obtained for $\Delta = 2n\pi$ ($n = 1, 2,\ldots$). Then, Eq. (11.19) renders

$$d_x = \frac{n\lambda}{2\sin\theta} \qquad (11.20)$$

Equation (11.20) is used for the determination of the in-plane displacement d_x. Similarly, we obtain the in-plane displacement d_y along the y-axis when the specimen is illuminated at equal angles with respect to the y-axis. Note that both in-plane displacements are independent of the out-of-plane displacement.

11.5 Shearography

Shearography or speckle pattern shearing interferometry is a full-field, non-contact, high-sensitivity interferometric method for measuring the first derivative of surface deformations. Shearography does not require a reference beam. The interference is generated by two identical laterally sheared object beams. The shearing effect can be realized by an optical wedge or a Michelson interferometer with tilted one of its mirrors. Since rigid-body movement does not create strain, shearography is insensitive to environmental vibrations and motion. The method was developed to overcome several limitations of holography by eliminating the reference beam.

A schematic arrangement of shearography is shown in Fig. 11.10. The specimen is illuminated by a collimated laser beam and the object image is captured by a camera. The shearing effect is created by placing a shearing device in front of the camera. The shearing device allows shearography to be a self-referenced interferometric system, as opposed to other interferometric methods.

Consider that the shearing effect is realized by a Michelson interferometer (Fig. 11.11). A point $P_1(x, y)$ on the object is mapped (due to the shearing effect imposed by tilting one mirror of the interferometer) to two points $P_1'(x, y)$, $P_1''(x +$

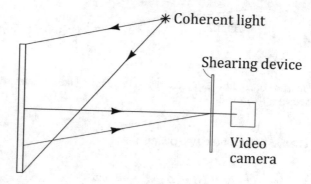

Fig. 11.10   A schematic arrangement of shearography

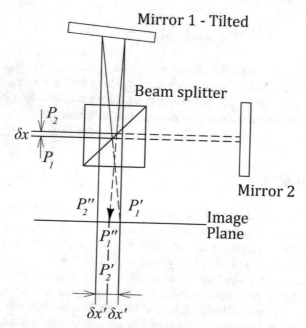

Fig. 11.11   Shearing effect produced by a Michelson interferometer

δx, y) on the image plane. Similarly, a second point $P_2(x + \delta x, y)$ at a distance δx from point $P_1(x, y)$ on the object is mapped to two points $P_2'(x + \delta x, y)$, $P_2''(x + 2\delta x, y)$ on the image plane. The rays $P_1''(x + \delta x, y)$ and $P_2'(x + \delta x, y)$ at point $(x + \delta x, y)$ interfere.

Let the waves from points $P_1(x, y)$ and $P_2(x + \delta x, y)$ be $a_1 e^{i\theta_1}$ and $a_2 e^{i\theta_2}$, respectively. The intensity I at point $(x + \delta x, y)$ is

$$I = \left(a_1 e^{i\theta_1} + a_2 e^{i\theta_2}\right)\left(a_1 e^{i\theta_1} + a_2 e^{i\theta_2}\right)^*$$
$$= a_1^2 + a_2^2 + 2a_1 a_2 \cos \varphi_{12} \tag{11.21}$$

where $\varphi_{12} = \theta_1 - \theta_2$.

When the object has deformed the waves from points $P_1(x, y)$ and $P_2(x + \delta x, y)$ will be $a_1 e^{(i\theta_1 + \Delta_1)}$ and $a_2 e^{(i\theta_2 + \Delta_2)}$, respectively. The intensity I' at point $(x + \delta x, y)$ is

$$I' = \left(a_1 e^{(i\theta_1 + \Delta_1)} + a_2 e^{(i\theta_2 + \Delta_2)}\right)\left(a_1 e^{(i\theta_1 + \Delta_1)} + a_2 e^{(i\theta_2 + \Delta_2)}\right)^*$$
$$= a_1^2 + a_2^2 + 2a_1 a_2 \cos(\varphi_{12} + \Delta_{12}) \tag{11.22}$$

where $\Delta_{12} = \Delta_1 - \Delta_2$.

For the interpretation of the interference fringes, we will use the sensitivity vector S given by Eq. (11.18). We have

$$\Delta_{12} = \Delta_1 - \Delta_2 = \frac{2\pi}{\lambda} \Delta r.S = \frac{2\pi}{\lambda}\left[(u_1 - u_2)i + (v_1 - v_2)j + (w_1 - w_2)k\right].S$$
$$= \frac{2\pi}{\lambda}[\delta u i + \delta v j + \delta w k].S \tag{11.23}$$

where u, v, w are the displacement components and i, j, k are the unit vectors along the x, y, z axes.

When the image is along the x-direction we have from Eq. (11.23)

$$\Delta_x = \delta x \frac{2\pi}{\lambda}\left[\frac{\delta u}{\delta x}i + \frac{\delta v}{\delta x}j + \frac{\delta w}{\delta x}k\right].S \tag{11.24}$$

Similarly, when the image is along the y-direction we have

$$\Delta_y = \delta y \frac{2\pi}{\lambda}\left[\frac{\delta u}{\delta y}i + \frac{\delta v}{\delta y}j + \frac{\delta w}{\delta y}k\right].S \tag{11.25}$$

Equations (11.24) and (11.25) are the general equations of shearography. Note that the phase difference is a function of the derivatives of displacement, unlike the other methods of coherent speckle interferometry in which the fringes represent contours of displacements.

Equations (11.24) and (11.25) can be simplified by suitably selecting the illumination and observation directions. When the direction of illumination lies on the x–z plane and the observation direction is along the z-axis (perpendicular to the plane of the specimen) the sensitivity vector S is given by Eq. (11.11). Thus we have

$$S.i = \sin\theta, \quad S.j = 0, \quad S.k = 1 + \cos\theta \tag{11.26}$$

and Eqs. (11.24) and (11.25) can be written as

$$\Delta_x = \frac{2\pi}{\lambda}\delta x\left[\sin\theta_{xz}\frac{\delta u}{\delta x} + (1 + \cos\theta_{xz})\frac{\delta w}{\delta x}\right] \tag{11.27}$$

$$\Delta_y = \frac{2\pi}{\lambda}\delta y\left[\sin\theta_{xz}\frac{\delta u}{\delta y} + (1 + \cos\theta_{xz})\frac{\delta w}{\delta y}\right] \tag{11.28}$$

Similarly, when the direction of illumination lies on the y–z plane we obtain

$$\Delta_x = \frac{2\pi}{\lambda}\delta x\left[\sin\theta_{yz}\frac{\delta u}{\delta x} + \left(1 + \cos\theta_{yz}\right)\frac{\delta w}{\delta x}\right] \tag{11.29}$$

$$\Delta_y = \frac{2\pi}{\lambda}\delta y\left[\sin\theta_{yz}\frac{\delta u}{\delta y} + \left(1 + \cos\theta_{yz}\right)\frac{\delta w}{\delta y}\right] \tag{11.30}$$

Equations (11.27) to (11.30) indicate that a shearogram contains in-plane and out-of-plane displacement derivatives. When the illumination direction is normal to the surface of the object the angle $\theta = 0$ ($\sin\theta = 0$, $\cos\theta = 1$), and Eqs. (11.27) and (11.28) give the derivatives of the out-of-plane displacement w as

$$\Delta_x = \frac{4\pi}{\lambda}\frac{\delta w}{\delta x}\delta x, \quad \Delta_y = \frac{4\pi}{\lambda}\frac{\delta w}{\delta y}\delta y \tag{11.31}$$

11.6 Electronic Speckle Pattern Interferometry (ESPI)

Electronic speckle pattern interferometry (ESPI) is the electronic version of speckle interferometry. Electronic devices (CCDs, Charge-Coupled Devices) are used to record and process the information. Because of the digital recording and processing ESPI is also named "digital speckle pattern interferometry". Television and computer processing systems are used to replace photographic recording materials and generate interferometric patterns. The object is imaged on a CCD camera by a lens system. The size of the speckle (typically of the order of 5–50 μm) is chosen to match the pixel size of the electronic target. The speckle patterns are recorded electronically and correlated numerically. The method does not require the strict stability conditions of speckle interferometry. The electronic version of shearography is called **digital shearography**. The method combines the advantages of shearography and digital recording and processing.

Further Readings

1. Khetan RP, Chiang FP (1976) Strain analysis by one-beam laser speckle interferometry. 1: single aperture method. Appl Opt 15:2205–2215
2. Dainty JC (1975) Laser speckle and related phenomena. Springer, Berlin
3. Jones R, Wykes C (1983) Holographic and speckle interferometry. Cambridge University Press, Cambridge
4. Sirohi BS (ed) (1993) Speckle metrology. CRC Press, Boca Raton
5. Briers JD (1997) Speckle techniques. In: Orr JF, Shelton JC (eds) Optical measurement methods in biomechanics. Chapman & Hall, UK, pp 76–98
6. Tyrer (1997) Electronic speckle pattern interferometry. In: Orr JF, Shelton JC (eds) Optical measurement methods in biomechanics. Chapman & Hall, UK, pp 99–124
7. Cloud GL (1998) Optical methods in engineering analysis. Cambridge University Press, Cambridge, pp 395–476
8. Gan Y, Steinchen W (2008) Speckle methods. In: Sharpe WN Jr (ed) Handbook of experimental solid mechanics. Springer, Berlin, pp 655–673
9. Sciammarella CA, Sciammarella FM (2012) Experimental mechanics of solids. Wiley, Hoboken, pp 547–606
10. Shukla A, Dally JW (2014) Experimental solid mechanics, 2nd edn. College House Enterprises, pp 415–438
11. Archbold E, Burch JM, Ennos AE (1970) Recording of in-plane surface displacements by double exposure speckle photography. Opt Acta 17:883–898
12. Archbold E, Ennos AE (1972) Displacement measurement from double-exposure laser photographs. Opt Acta 19:253–271
13. Duffy DE (1974) Measurement of surface displacement normal to the line of sight. Exp Mech 14:378–384
14. Hung YY, Taylor CR (1974) Measurement of slopes of structural deflections by speckle-shearing interferometry. Exp Mech 14:281–285
15. Stetson KA (1975) A review of speckle photography and interferometry. Opt Eng 14:482–489
16. Cloud G (1975) Practical speckle interferometry for measuring in-plane deformation. Appl Opt 14:878–884
17. Chiang FP (1979) Optical stress analysis using moiré fringe and laser speckles. Opt Eng 18:448–455
18. Parks VJ (1980) The range of speckle metrology. Exp Mech 20:181–191
19. Jacquot P (2008) Speckle interferometry: a review of the principal methods in use for experimental mechanics applications. Strain 44:57–69
20. De la Torre IM, Socorro Hernández Montes M, Mauricio Flores-Moreno J, Santoyo FM (2016) Laser speckle based digital optical methods in structural mechanics: a review. Opt Lasers Eng 87:32–58
21. Yang L, Li J (2018) Shearography. In: Ida N, Mayendorf N (eds) Handbook of advances in nondestructive evaluation. Springer, Berlin, pp 2–37

Chapter 12
Digital Image Correlation (DIC)

12.1 Introduction

Digital image correlation (DIC) is a full-field non-contacting optical method that can capture the shape, motion, and deformation of solid objects. The basis of the method is the matching of the gray values of points from an image of the surface of an object before and after deformation. The gray values of points of images of an object are acquired, stored, digitized, and correlated (matched) to compute shape and surface displacements. A matching process based on gray intensity levels is performed, hence the name of the method *digital image correlation*. DIC techniques can be applied to macro-, micro-, and nano-scale mechanical testing under static and dynamic loading. The development of DIC is due to the advances in computer technology and digital cameras.

DIC methods use incoherent light for the illumination and there is no need for vibration isolated table and optical components, like prisms, filters, beam splitters, as in other optical methods for displacement measurement, like holography, interferometry, and speckle methods. For the application of DIC methods for surface deformation measurements, a high-resolution digital camera and computer software are mainly required. Today, DIC methods are the most widely used experimental methods in solid mechanics for deformation measurements.

In the following, we will present the essential steps of DIC, the speckle patterning, the image digitization process, the intensity interpolation, and the correlation of images for measurement of two-, three-dimension and volumetric displacements.

© The Author(s), under exclusive license to Springer Nature Switzerland AG 2022   251
E. E. Gdoutos, *Experimental Mechanics*, Solid Mechanics and Its Applications 269,
https://doi.org/10.1007/978-3-030-89466-5_12

12.2   Steps of DIC

Application of DIC consists of the following steps:

1.   A high-contrast tracking pattern often called *speckle pattern* is installed on the surface of the object. It may consist of dots, lines, regular or random arrays. It is assumed that there is a one-to-one correspondence between the motions of points in the image and the motions of points on the object.
2.   A portion of the pattern, called a *subset*, is selected for tracking. It is assumed that each subset has an adequate variation of light intensity.
3.   Images of the object are obtained before and after deformation. The image before deformation is defined as the *reference image* and the image after deformation as the *deformed image*.
4.   The reference and deformed images are digitized and stored in a computer.
5.   The digitized images are cross-correlated (matched) to calculate the displacements between the reference and the deformed images. This is accomplished with a matching process based on gray intensity levels.

In the following, we discuss the above five steps in more detail.

12.3   Speckle Patterning

To define accurate matching between images, the object surface must have high contrast (adequate light intensity variations) as a carrier of deformation information. Occasionally the surface of the object may have features that suffice for a natural speckle pattern, like lines, grids, dots. Usually, stochastic patterns are applied to the surface of the object. The quality of the DIC results depends on the quality of the speckle pattern. The pattern must remain attached to the surface of the object. Several approaches have been developed to achieve a high-contrast speckle pattern of the required size. Some general features of a successful speckle pattern are:

1.   The speckles are random in position but uniform in size.
2.   The size of the speckle is at least *3-by-3* pixels to avoid aliasing, but not more than *7-by-7* pixels to achieve a relatively high density of speckles. For speckles larger than *7-by-7* pixels there will be few data points for analysis.
3.   The speckle pattern needs to have good grayscale contrast. This can be visualized by a histogram with the number of pixels plotted with respect to the grayscale level. A broad flat histogram over the range of the intensity pattern is optimal.
4.   The pattern must have a speckle density of about 50%.
5.   The pattern must remain stable and not detach from the surface in the testing environment (e.g., high-temperature experiments).

Random artificial speckle patterns are generated by several methods including paint, inks and dyes, powder particles, nanoparticles, lithographed particles. High-quality speckle patterns are applied quickly with spraying paint. Painted patterns are compliant with most engineering materials. Black and white paints provide the best contrast. White paint as the background and black as the speckles is preferred over the converse because black paint maintains better contrast over white paint. For materials with high deformations, inks and dyes are preferable over paint since the latter is not stretchy enough. Powder particles are preferable in sticky materials since they better adhere than paint and produce smaller speckles than painted patterns. Nanoparticles for applications with scanning electron microscopy can produce smaller particles (of about 20–100 nm) than powders. Small speckle patterns can also be achieved by lithography with higher degree of control than most of the microscale patterning methods.

12.4   Image Digitization

Good images of objects under study are a key step in data collection for DIC. Appropriate image magnification depends on the length scale of the samples. Good photography or microscopy translate to good images. A single camera for two-dimensional analysis or two synchronized cameras for three-dimensional analysis is used. Typically, scientific-grade digital cameras are used to obtain high-quality images on the sensor plane. Both CCD (charge-coupled device) and CMOS (metal–oxide–semiconductor) cameras have been successfully used in a wide range of image applications. Typical cameras have 2048×2048 pixels in the sensor array and 8–10 bit (8 bits gives 2^8 or 256 gray levels, 10 bits gives 2^{10} or 1024 gray levels, etc.) intensity resolution levels for each pixel. A typical pixel size is six microns. The incident radiation on the camera is converted into an integer value with N bits of intensity resolution.

The digitized image intensity field is the main experimental data for DIC analysis. The digitized intensity values are stored in a computer with digital image acquisition components.

12.5   Intensity Interpolation

The intensity values or gray levels of a digital image are only obtained at positions of the camera sensors or pixels. Pixel locations in the reference state generally move to non-pixel locations in the deformed state of the object. To compare the reference and deformed images corresponding to the unloaded and loaded positions of the object intensity values at pixel locations need to be compared with intensity values at non-pixel locations. Therefore, the discretized intensity pattern should be converted to a continuous intensity pattern. Various interpolation methods, e.g., bilinear, bi-cubic, bi-cubic spline, etc. can be used. Let us consider four pixels with coordinated (0, 0), (1, 0), (0, 1) (1, 1) in a local coordinate system Ox_1x_2 (Fig. 12.1). For a bilinear interpolation, the intensity function $I(x_1, x_2)$ can be expressed by

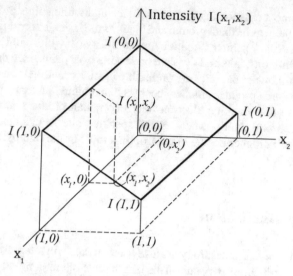

Fig. 12.1 Bilinear interpolation of the intensity distribution between four points of a subregion

$$I(x_1, x_2) = ax_1 + bx_2 + cx_1x_2 + d \tag{12.1}$$

where

$$
\begin{aligned}
a &= I(1, 0) - I(0, 0) \\
b &= I(0, 1) - I(0, 0) \\
c &= I(1, 1) - I(0, 0) - I(0, 1) - I(1, 0) \\
d &= I(0, 0)
\end{aligned}
\tag{12.2}
$$

For a bi-cubic spline interpolation the intensity function $I(x_1, x_2)$ can be expressed by

$$I(x_1, x_2) = \sum_{m=1, n=1}^{m=3, n=3} a_{mn} x_1^m x_n^2 \tag{12.3}$$

where a_{mn} are fitting coefficients. For the determination of the nine coefficients $a_{mn}(m, n = 1, 2, 3)$ at least a 3×3 pixel array of intensity data is required.

12.6 Image Correlation—Displacement Measurement

Consider a point $P(x_1, x_2)$ in the reference image which is mapped into point $p(x_1^*, x_2^*)$ in the deformed image (Fig. 12.2). If (u_1, u_2) is the displacement vector we obtain

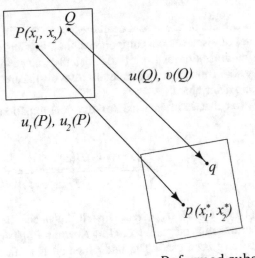

Reference subset

Deformed subset

Fig. 12.2 Point $P(x_1, x_2)$ in the reference image is mapped into point $p(x_1^*, x_2^*)$ in the deformed image

by approximating the displacement components by a second order Taylor series expansion

$$x_1^* = x_1 + u_1 + \frac{\partial u_1}{\partial x_1}\Delta x_1 + \frac{\partial u_1}{\partial x_2}\Delta x_2 + \frac{1}{2}\frac{\partial^2 u_1}{\partial x_1^2}\Delta x_1^2 + \frac{\partial^2 u_1}{\partial^2 x_1 x_2} + \frac{1}{2}\frac{\partial^2 u_1}{\partial x_2^2}\Delta x_2^2$$

$$x_2^* = x_2 + u_2 + \frac{\partial u_2}{\partial x_1}\Delta x_1 + \frac{\partial u_2}{\partial x_2}\Delta x_2 + \frac{1}{2}\frac{\partial^2 u_2}{\partial x_1^2}\Delta x_1^2 + \frac{\partial^2 u_2}{\partial^2 x_1 x_2} + \frac{1}{2}\frac{\partial^2 u_2}{\partial x_2^2}\Delta x_2^2$$

$$(12.4)$$

Equation (12.4) indicates that the coordinates x_1^*, x_2^* of point p are obtained from the coordinates x_1, x_2 of point P and twelve parameters. They are: the displacement components u_1 and u_2, the first derivatives of displacement components $\frac{\partial u_1}{\partial x_1}, \frac{\partial u_1}{\partial x_2}, \frac{\partial u_2}{\partial x_1}, \frac{\partial u_2}{\partial x_2}$ and the second derivatives of displacement components $\frac{\partial^2 u_1}{\partial x_1^2}, \frac{\partial^2 u_1}{\partial^2 x_1 x_2}, \frac{\partial^2 u_1}{\partial x_2^2}, \frac{\partial^2 u_2}{\partial x_1^2}, \frac{\partial^2 u_2}{\partial^2 x_1 x_2}, \frac{\partial^2 u_2}{\partial x_2^2}$.

Let $I(x_1, x_2)$ be the light intensity at point $P(x_1, x_2)$ and $I'(x_1^*, x_2^*)$ the light intensity at point $p(x_1^*, x_2^*)$. Several mathematical expressions can be used to correlate the intensities $I(x_1, x_2)$ and $I'(x_1^*, x_2^*)$. A least square correlation can be written in the form

$$C\left(u_1, u_2, \frac{\partial u_1}{\partial x_1}, \frac{\partial u_1}{\partial x_2}, \frac{\partial u_2}{\partial x_1}, \frac{\partial u_2}{\partial x_2}, \frac{\partial^2 u_1}{\partial x_1^2}, \frac{\partial^2 u_1}{\partial^2 x_1 x_2}, \frac{\partial^2 u_1}{\partial x_2^2}, \frac{\partial^2 u_2}{\partial x_1^2}, \frac{\partial^2 u_2}{\partial^2 x_1 x_2}, \frac{\partial^2 u_2}{\partial x_2^2}\right)$$

$$= \sum \left[I'(x_1^*, x_2^*) - I(x_1, x_2) \right]^2 \tag{12.5}$$

where the summation takes on the subset of interest from the digital image.

The twelve values of displacement components, their first and second derivatives are calculated by minimization of Eq. (12.5). Usually, the second derivatives are neglected and only six displacement parameters (two displacements and four first-order derivatives) are determined.

Another expression that is often used for the determination of the unknown displacements is

$$1 - C = 1 - \frac{\sum \left[I(x_1, x_2) I'(x_1^*, x_2^*) \right]}{\sqrt{\sum \left[I^2(x_1, x_2) I'^2(x_1^*, x_2^*) \right]}} \tag{12.6}$$

Note that if $(1 - C) = 0$ perfect match between the two images is achieved. Several optimization methods including Newton–Raphson and Levenberg–Marquardt methods have been used. The latter method is as fast as the Newton–Raphson and has better convergence. The magnitude of $(1 - C)$ is an indicator of the accuracy of determining the displacements. Values of $(1 - C) < 0.001$ are considered as good match.

12.7 2-D DIC

For 2-D displacement measurements, one camera suffices. The application of the method is simple. A noncoherent light source and a high-resolution digital camera are all that are needed. The surface of the 2-D object needs to be flat and remain flat during the experiment and be parallel to the sensor array on the image plane of the camera. Furthermore, the out-of-plane displacement $u_3 = u_3(x_1, x_2)$ of the surface relative to the camera needs to be small. For a strain error smaller than $\Delta \varepsilon$ the following equation must be satisfied

$$\frac{u_3(x_1, x_2)}{z} < \Delta \varepsilon \tag{12.7}$$

where z is the distance between the point (x, y) on the surface and the lens.

For the calibration of the camera-body system, a correspondence between points on the object and points on the image can be established by the following equations

$$X_s = \Lambda_x X + C_x$$
$$Y_s = \Lambda_y Y + C_y \tag{12.8}$$

where (X, Y) is position on the object, (X_s, Y_s) is position on the image in terms of pixels, Λ_x and Λ_y are magnification factors with dimensions pixels per unit length on the object, (C_x, C_y) is the pixel location of the image center point (for a 1024×1024 sensor array $C_x = C_y = 512$).

For small distortions, calibration for C_x, C_y is not required, since they are constants and do not affect displacements or strains. Values of Λ_x and Λ_y can be obtained by imaging a ruler or a grid with alignment along the row and column directions in the sensor plane and analyzing the images to obtain pixel positions of several of the marks having a known spacing along both the horizontal and row directions.

12.8 3-D DIC

As it was discussed in 2-D DIC, single camera systems are limited to plane specimens. To capture the 3-D displacement components of a body two or more cameras observing the surface from different directions are needed. Each camera has to be calibrated either separately or the camera system has to be calibrated. Following the calibration of the cameras, image acquisition is the same in both 2-D and 3-D image correlation.

Both 2-D and 3-D DIC methods are easy to apply. Ordinary white light is used for illuminating the surface of the specimen. The methods are relatively insensitive to vibrations and a vibration isolated table is not necessary, although the body and the camera need to be stable during the course of the experiment.

12.9 Volumetric Digital Image Correlation (V-DIC)

The concepts of 2-D DIC can be extended to make measurements of displacements inside bodies. These methods are known as volumetric digital image correlation methods (V-DIC). Internal images of the body can be obtained using Computed Tomography (CT), Magnetic Resonance Imaging (MRI), Positron Emission Tomography (PET). The internal images are correlated in the same way as in 2-D DIC.

Further Readings

1. Gonzalez RC, Woods RE (1987) Digital image processing, 2nd edn. Addison Wesley, Boston
2. Sutton MA (2008) Digital image correlation for shape and deformation measurements. In: Sharpe WN Jr (ed) Springer handbook of experimental solid mechanics. Springer, Berlin, pp 565–600
3. Sutton MA, Orteu JJ, Schreier HW (2009) Image correlation for shape, motion and deformation measurements. Springer, Berlin
4. Sciammarella CA, Sciammarella FM (2012) Experimental mechanics of solids. Wiley, Hoboken, pp 607–629

5. Shukla A, Dally JW (2014) Experimental solid mechanics, 2nd edn. College House Enterprises, pp 439–474
6. Jahne B (2017) Digital image processing-concepts, algorithms and scientific applications. Springer, Berlin
7. Niezrecki C, Baqersad J, Sabato A (2018) Digital image correlation techniques for NDE and SHM. In: Ida, N, Meyendorf N (eds) Handbook of advanced non-destructive evaluation. Springer, Berlin, pp 1–46
8. Lamberti, L, Lin M-T, Furlog C, Sciammarella C, Reu PL, Sutton MA (eds) (2019) Advancement of optical methods & digital image correlation in experimental mechanics, vol 3. Springer, Berlin
9. Périé J-N, Passieux J-C (eds) (2020) Advances in digital image correlation (DIC). Mdpi AG, Basel
10. Lin M-T, Sciammarella C, Espinosa H, Furlog C, Lamberti, L, Reu PL, Sutton MA, Hwang CH (eds) (2020) Advancements in optical methods & digital image correlation in experimental mechanics, vol 3. Springer, Berlin
11. Lin M-T, Cosme Furlong C, Kuo-Cheng Huang K-C (eds) (2021) Advancements of optical methods & digital image correlation in experimental mechanics. Springer, Berlin
12. Peters WH, Ranson WF (1982) Digital imaging techniques in experimental mechanics. Opt Eng 21:427–431
13. Sutton MA, Wolters WJ, Peters WH, Ranson WF, McNeill SR (1983) Determination of displacements using an improved digital correlation method. Image Vis Comput 1:133–139
14. Peters WH, Zheng-Hui H, Sutton MA, Ranson WF (1984) Two-dimensional fluid velocity measurements by use of digital speckle correlation techniques. Exp Mech 24:117–121
15. Anderson J, Peters WH, Sutton MA, Ranson WF, Chu TC (1984) Application of digital correlation methods to rigid body mechanics. Opt Eng 22:738–742
16. Chu TC, Ranson WF, Sutton MA, Peters WH (1985) Applications of digital image correlation techniques to experimental mechanics. Exp Mech 25:232–245
17. Sutton MA, McNeill SR, Jang J, Babai M (1988) The effects of subpixel image restoration on digital correlation error estimates. Opt Eng 10:870–877
18. Sutton MA, Cheng M, McNeill SR, Chao YJ, Peters WH (1988) Application of an optimized digital correlation method to planar deformation analysis. Image Vis Comput 4(3):143–150
19. Bruck HA, McNeill SR, Sutton MA, Peters WH III (1989) Digital image correlation using Newton-Raphson method of partial differential correction. Exp Mech 29:261–267
20. Sutton MA, Turner JL, Chae TL, Bruck HA (1991) Full field representation of discretely sampled surface deformation for displacement and strain analysis. Exp Mech 31(2):168–177
21. Sutton MA, Bruck HA, Chae TL, Turner JL (1991) Experimental investigations of three-dimensional effects near a crack tip using computer vision. Int J Fract 53:201–228
22. Han G, Sutton MA, Chao YJ (1994) A study of stationary crack tip deformation fields in thin sheets by computer vision. Exp Mech 34(4):357–369
23. Han G, Sutton MA, Chao YJ (1995) A study of stable crack growth in thin SEC specimens of 304 stainless steel. Eng Fract Mech 52(3):525–555
24. Liu J, Sutton MA, Lyons JS (1998) Experimental characterization of crack tip deformations in alloy 718 at high temperatures. ASME J Eng Mater Technol 20(1):71–78
25. Sutton MA, McNeill SR, Helm JD, Chao YJ (2000) Advances in two-dimensional and three-dimensional computer vision. In: Rastogi PK (ed) Photomechanics. Topics Appl Phys 77:323–372
26. Schreier HW, Braasch JR, Sutton MA (2000) Systematic errors in digital image correlation caused by intensity interpolation. Opt Eng 39(11):2915–2921
27. Schreier HW, Sutton MA (2002) Systematic errors in digital image correlation due to undermatched subset shape functions. Exp Mech 42:303–310
28. Sutton A, Yan JH, Tiwaria V, Schreierb HW, Orteu JJ (2008) The effect of out-of-plane motion on 2D and 3D digital image correlation measurements. Opt Lasers Eng 46:746–757
29. Reu P (2012) The art and application of DIC. Stereo-rig design: camera selection - Part 2. Exp Tech 36:3–4

30. Reu P (2012) The art and application of DIC. Hidden components of DIC: calibration and shape function - Part 1. Exp Tech 36:3–5
31. Reu P (2012) The art and application of DIC. Hidden components of 3D-DIC: interpolation and matching - Part 2. Exp Tech 36:3–4
32. Reu P (2012) The art and application of DIC. Hidden components of 3D-DIC: triangulation and post-processing - Part 3. Exp Tech 36:3–5
33. Reu P (2012) The art and application of DIC. Stereo-rig design: creating the stereo-rig layout - Part 1. Exp Tech 36:3–4
34. Reu P (2013) The art and application of DIC. Stereo-rig design: lens selection - Part 3. Exp Tech 37:1–3
35. Reu P (2013) The art and application of DIC. Stereo-rig design: stereo-angle selection - Part 4. Exp Tech 37:1–2
36. Reu P (2013) The art and application of DIC. Stereo-rig design: lighting - Part 5. Exp Tech 37:1–2
37. Reu P (2013) The art and application of DIC. Calibration: pre-calibration routines. Exp Tech 37:1–2
38. Reu P (2013) The art and application of DIC. Calibration: 2D calibration. Exp Tech 37:1–2
39. Reu P (2013) The art and application of DIC. Calibration: a good calibration image. Exp Tech 37:1–3
40. Reu P (2014) The art and application of DIC. Calibration: stereo calibration. Exp Tech 38:1–2
41. Reu P (2014) The art and application of DIC. Calibration: sanity checks. Exp Tech 38:1–2
42. Reu P (2014) The art and application of DIC. Calibration: care and feeding of a stereo-rig. Exp Tech 38:1–2
43. Reu P (2014) Speckles and their relationship to the digital camera. Exp Tech 38:1–2
44. Reu P (2014) All about speckles: aliasing. Exp Tech 38:1–3
45. Reu P (2014) All about speckles: speckle size measurement. Exp Tech 38:1–2
46. Pan B (2018) Digital image correlation for surface deformation measurement: historical developments, recent advances and future goals. Meas Sci Technol 29:1–3

Chapter 13
Thermoelastic Stress Analysis (TSA)

13.1 Introduction

Thermoelastic stress analysis (TSA) is a full-field non-contact optical method for measuring the stresses on the surface of bodies. The method is based on the thermoelastic effect according to which when a material is subjected to a cyclical load a temperature variation is produced. Under adiabatic conditions, the temperature variation for isotropic materials is proportional to the sum of the two surface principal stresses. The temperature changes are very small of the order of 0.001 °C. The theoretical basis of the thermoelastic effect is known for more than 150 years. However, TSA appeared in the last 30 years due to the advent of infrared (IR) detectors capable to monitor very small temperature changes.

The method involves little setup and specimen preparation. The sensitivity of the method is similar to that of strain gages. A disadvantage of the method is the relatively high cost of commercial TSA systems, even though hardware affordability has been greatly improved over the past years.

In the following, we present the theoretical basis of the thermoelastic effect, the infrared detectors for measuring very small temperature changes, the conditions of adiabaticity, the specimen preparation, the stress separation methods, and applications of TSA.

13.2 Thermoelastic Law

The thermoelastic law of solids is analogous to that of gases. When a gas is compressed its temperature increases and when it is expanded its temperature decreases. The same happens in solids, an elastic solid subjected to a compressive load will experience a temperature increase, while when it is subjected to a tensile load will experience a temperature decrease. The difference between solids and gases is that the temperature change is orders of magnitude smaller in solids, of

© The Author(s), under exclusive license to Springer Nature Switzerland AG 2022
E. E. Gdoutos, *Experimental Mechanics*, Solid Mechanics and Its Applications 269,
https://doi.org/10.1007/978-3-030-89466-5_13

the order of milli-Celsius than in gases. The cyclic variation of temperature produces a cyclic variation of the photon emission from the surface of the material, which can be measured by appropriate sensors. The load change and the temperature change are related to the thermoelastic law developed by Lord Kelvin in 1853.

According to thermoelastic law, the change of temperature ΔT and the change of strain $\Delta \varepsilon_{ij}$ are related by

$$\Delta T = \frac{T}{\rho C_E} \frac{\partial \sigma_{ij}}{\partial T} \Delta \varepsilon_{ij} + \frac{Q}{\rho C_E} \qquad (13.1)$$

where

T the absolute temperature of the material $[T(K) = T(°C) + 273.16]$.
C_E the specific heat at constant strain.
ρ the mass density.
σ_{ij} the stress tensor.
ε_{ij} the strain tensor.
Q the heat input.

In Eq. (13.1) the indicial notation is used where summation is implied on an index that appears twice $(i, j = 1, 2)$.

TSA experiments are conducted under adiabatic conditions (no heat conduction takes place). Equation (13.1) with $Q = 0$ becomes

$$\Delta T = \frac{T}{\rho C_E} \frac{\partial \sigma_{ij}}{\partial T} \Delta \varepsilon_{ij} \qquad (13.2)$$

The term $\partial \sigma_{ij}/\partial T \Delta \varepsilon_{ij}$ for conditions of plane stress $(\sigma_{13} = \sigma_{23} = \sigma_{33} = 0)$ is given by

$$\frac{\partial \sigma_{ij}}{\partial T} \Delta \varepsilon_{ij} = \frac{\partial \sigma_{11}}{\partial T} \Delta \varepsilon_{11} + \frac{\partial \sigma_{22}}{\partial T} \Delta \varepsilon_{22} + \frac{\partial \sigma_{12}}{\partial T} \Delta \varepsilon_{12} \qquad (13.3)$$

The linear elastic stress–strain-temperature law for isotropic materials is

$$\sigma_{11} = \frac{E}{1 - v^2}(\varepsilon_{11} + v\varepsilon_{22}) - \frac{E\alpha}{1 - v}T$$

$$\sigma_{22} = \frac{E}{1 - v^2}(\varepsilon_{22} + v\varepsilon_{11}) - \frac{E\alpha}{1 - v}T$$

$$\sigma_{11} = \frac{E}{1 + v}\varepsilon_{12} \qquad (13.4)$$

where α is the coefficient of linear thermal expansion.

Differentiating Eq. (13.4) with respect to temperature T we obtain

$$\frac{\partial \sigma_{11}}{\partial T} = -\frac{E\alpha}{1 - v}$$

$$\frac{\partial \sigma_{22}}{\partial T} = -\frac{E\alpha}{1-\nu}$$

$$\frac{\partial \sigma_{12}}{\partial T} = 0 \tag{13.5}$$

The last equation indicates that the thermoelastic effect is independent of the shear stress. From Eqs. (13.3) and (13.5) we obtain

$$\frac{\partial \sigma_{ij}}{\partial T} \Delta \varepsilon_{ij} = -\frac{E\alpha}{1-\nu} \Delta(\varepsilon_{11} + \varepsilon_{22}) \tag{13.6}$$

Eq. (13.4) renders

$$\Delta(\varepsilon_{11} + \varepsilon_{22}) = \frac{1-\nu}{E} \Delta(\sigma_{11} + \sigma_{22}) + 2\alpha \Delta T \tag{13.7}$$

From Eqs. (13.2), (13.6) and (13.7) we obtain

$$\Delta T = -\frac{T}{\rho C_E} \left[\alpha \Delta(\sigma_{11} + \sigma_{22}) + \frac{2\alpha^2 E \Delta T}{1-\nu} \right] \tag{13.8}$$

Eq. (13.8) renders

$$\Delta(\sigma_{11} + \sigma_{22}) = -\frac{\Delta T}{\alpha} \left[\frac{\rho C_E}{T} + \frac{2E\alpha^2}{1-\nu} \right] \tag{13.9}$$

Let us now introduce a new constant, the specific heat at constant pressure C_P. It is related to the specific heat at constant strain C_E introduced previously by

$$C_P = C_E + \frac{2E\alpha^2 T}{\rho(1-\nu)} \tag{13.10}$$

Then, Eq. (13.9) becomes

$$\Delta(\sigma_{11} + \sigma_{22}) = -\frac{\Delta T}{\alpha} C_P \frac{\rho}{T} \tag{13.11}$$

Introducing a new constant K by

$$K = \frac{\alpha}{\rho C_P} \tag{13.12}$$

Equation (13.11) takes the form

$$\Delta T = -KT\Delta(\sigma_{11} + \sigma_{22}) \tag{13.13}$$

Table 13.1 Thermoelastic response of several engineering materials

Material	a_T (10^{-6} K^{-1})	ρ (kg m^{-3})	C_p (J kg^{-1} K^{-1})	$\delta T/\delta\sigma$ (mK MPa^{-1})
Al alloy (Al2024)	22.8	2770	875	−2.76
Ti alloy (Ti6A14V)	8.6	4430	526	−1.09
Steel (AISI1005)	12.6	7872	481	−0.98

Equation (13.13) **is the fundamental equation of TSA**. It relates two quantities: the change of temperature ΔT and the change of the sum of the principal stresses $\Delta(\sigma_{11} + \sigma_{22})$ through a material constant K and the absolute temperature T. The material constant K is referred to as **thermoelastic constant**. Note that the relation between ΔT and $\Delta(\sigma_{11} + \sigma_{22})$ is linear.

Equation (13.13) indicates: firstly, that pure shear does not produce thermoelastic effect, secondly, a positive change in stress produces a negative change in temperature [the minus sign in Eq. (13.13)]. The thermoelastic effect is a reversible phenomenon, that is, under adiabatic conditions, a material deformed and then released recovers to its initial thermal state. Adiabatic conditions are never attained in practice. Adiabaticity is improved by increasing the loading rate. At static loading, the thermoelastic response is extinguished by diffusion.

The temperature variations produced by the thermoelastic effect are small. Table 13.1 [1] presents the thermoelastic response of three engineering materials. Note that at the elastic limit of Al and Ti alloys the temperature variation approaches 1 K. For example, for an Al alloy a stress change of 10 MPa, which corresponds to a strain change of 140 με, creates a temperature change of 27.6 mK.

Equation (13.13) was derived for isotropic materials. For orthotropic materials, the stress term is replaced by $(K_1\Delta\sigma_{11} + K_2\Delta\sigma_{22})$ where K_1 and K_2 are thermomechanical coefficients and σ_{11} and σ_{22} are the stresses in the directions of material symmetry.

13.3 Infrared Detectors

Even though the thermoelastic effect was discovered in 1853, TSA appeared a few decades ago. Temperature measurements of the order of milli Kelvin produced by the thermoelastic effect cannot be achieved without the advent of infrared (IR) detectors with this sensitivity. Infrared radiation (IR) is electromagnetic radiation with wavelengths longer than the wavelength of the nominal red edge of the visible spectrum of $\lambda = 700$ nm. It extends up to $\lambda = 1$ mm. IR is invisible by the human eye. Most of the thermal radiation emitted by objects is IR. IR behaves both as a wave and as a particle, the photon. IR detectors are transducers converting the energy of the incident photons into voltage. They are categorized as quantum and thermal/nonquantum detectors.

Quantum detectors are based on the photoelectric effect, according to which when light (photons) strike surface electrons are emitted. For the emission of electrons, the frequency of the light should be greater than some minimum value called the threshold frequency. For a detector produced from indium-antimonide, a photon of wavelength 5.6 μm produces 0.22 eV. Photons with longer wavelengths are undetected. The quantum detectors need to be cooled at about −200 °C. The first commercial infrared camera named SPATE (acronym for Stress Pattern Analysis by Thermal Emission) used quantum IR detectors.

Thermal/nonquantum detectors called **bolometers** detect the absorption of heat rather than photons and transform it into an electrical signal. Bolometers, known as thermal detectors, measure the power of the incident radiation by heating a material with a temperature-dependent electrical resistance. They do not need to be cooled, are more compact in size, are more tolerant to vibrations and they have a considerably lower cost than the quantum detectors.

13.4 Adiabaticity

The fundamental Eq. (13.13) of TSA was derived from the thermodynamic Eq. (13.1) by putting $Q = 0$, that is, assuming adiabatic conditions. Heat transfer in a specimen depends on the thermal conductivity, the temperature gradients, and the frequency of the applied load. The only condition which can be controlled for adiabatic conditions to prevail is the load frequency. It has been established that adiabatic behavior occurs for loading frequencies above a threshold value. For most metals, this threshold value is 2 Hz. However, higher loading frequencies improve the condition of adiabaticity. Frequencies in the range of 10–25 Hz appear suitable for aluminum and graphite- and glass–epoxy composites. The condition of adiabaticity can be checked by comparing the phase of the response signal with the phase of the load signal. A measurable phase shift will occur if heat conduction takes place.

13.5 Specimen Preparation

The change of surface temperature of a body subjected to change of surface stress is greatly increased if the surface has a high emissivity in the IR radiation. To enhance the surface emissivity specimens are covered with a thin coat of matt black paint. This coat ensures that the surface of the specimen approaches that of *black body* with an emissivity coefficient of unity. Besides that, the coat ensures the uniformity of the emissivity of the specimen. Coat thicknesses of the order of 20–30 μm are usually used with load frequency in the range of 5–200 Hz. The temperature change of the surface of the coating should be the same as that on the surface of the specimen. The coating needs to be both thin and uniform.

13.6   Calibration

Equation (13.13) can be directly applied for the determination of the change of the sum of the two principal stresses (for isotropic material under plane stress conditions) from the temperature change when the material constant K is known. Constant K is given by Eq. (13.12) in terms of three material parameters. Furthermore, conversion of temperature changes to electrical signals introduces further parameters. In this respect, it is more convenient to use a calibration test for the determination of K. Calibration specimens are of the same material and painting coating and should be tested under the same loading frequency and environmental conditions, as the test body. Calibration methods include those in which the state of stress in a specimen is known (such as the tension specimen, beam under four-point bending, circular disk under diametral compression) and those in which the state of stress at a point of the specimen is determined by strain gages, finite elements, etc.

13.7   Stress Separation

TSA is a full-field optical method for the determination of isopachics (contours of equal sum of the principal stresses $(\sigma_1 + \sigma_2)$ for isotropic materials). For the determination of the individual stresses, σ_1 and σ_2 complementary methods are needed. TSA can be combined with photoelasticity from which the difference of the principal stresses $(\sigma_1 - \sigma_2)$ is determined. Other methods include finite/boundary elements, DIC, etc. Field equations of solid mechanics, real and complex stress functions, and other methods have also been used for stress separation.

13.8   Applications

TSA is a full-field non-contact optical method. It has been applied to many problems including composite materials, fracture mechanics and fatigue, residual stresses, vibration analysis, among others. Field applications of the method include non-destructive testing and structural health monitoring cases.

Further Readings

1. Rajic N, Rowlands D (2013) Thermoelastic stress analysis with a compact low-cost microbolometer system. Quant Infrared Thermography 10(2):135–158
2. Rauch BJ, Rowlands RE (1993) Thermoelastic stress analysis. In: Kobayashi AS (ed) Handbook of experimental mechanics, 2nd edn. Society for Experimental Mechanics, Bethel, pp 581–599

3. Greene RJ, Patterson EA, Rowlands RE (2008) Thermoelastic stress analysis. In: Sharpe WN Jr (ed) Handbook of experimental solid mechanics. Springer, Berlin, pp 743–767
4. Backman D (2018) Thermoelastic stress analysis. In: ASM handbook, vol 17, pp 143–151
5. Ryall TG, Wong AK (1988) Determining stress components from thermoelastic data—a theoretical study. Mech Mater 7:205–214
6. Stanley P (1989) Stress separation from SPATE data for a rotationally symmetrical pressure vessel. Proc SPIE 108(4):72–83
7. Enke NF (1989) An enhanced theory for thermoelastic stress analysis of isotropic materials. SPIE 1084:84–102
8. Huang YM, Abdel Mohsen H, Rowlands RE (1990) Determination of individual stresses thermoelastically. Exp Mech 30(1):88–94
9. Huang YM, Rowlands RE, Lesniak JR (1990) Simultaneous stress separation, smoothing of measured thermoelastic information, and enhanced boundary data. Exp Mech 30:398–403
10. Stanley P (1997) Applications and potential of thermoelastic stress analysis. J Mat Proc Tech 64:359–370
11. Dulieu-Barton JM, Stanley P (1997) Reproducibility and reliability of the response from four SPATE systems. Exp Mech 37(4):440–444
12. Dulieu-Barton JM, Stanley P (1998) Development and applications of thermoelastic stress analysis. J Strain Analysis Eng Des 33:93–104
13. Dulieu-Smith JM, Stanley P (1998) On the interpretation and significance of the Grüneisen parameter in thermoelastic stress analysis. J Mat Proc Technol 75:75–83
14. Dulieu-Barton JM (1999) Introduction to thermoelastic stress analysis. Strain 35(2):35–39
15. Dulieu-Barton JM, Stanley P (1999) Applications of thermoelastic stress analysis to composite materials. Strain 35(2):41–48
16. Tomlinson RA, Olden EJ (1999) Thermoelasticity for the analysis of crack tip stress fields—a review. Strain 35:49–55
17. Pitarresi G, Patterson EA (2003) A review of the general theory of thermoelastic stress analysis. J Strain Analysis Eng Des 38:405–417
18. Tomlinson RA, Marsavina L (2004) Thermoelastic investigations for fatigue life assessment. Exp Mech 44:487–494
19. Diaz FA, Yates JR, Patterson EA (2004) Some improvements in the analysis of fatigue cracks using thermoelasticity. Int J Fatigue 26:365–376
20. Quinn S, Dulieu-Barton JM, Langlands JM (2004) Progress in thermoelastic residual stress measurement. Strain 40:127–133
21. Stanley P (2008) Beginnings and early development of thermoelastic stress analysis. Strain 44:285–297
22. Rajic N, Galea S (2015) Thermoelastic stress analysis and structural health monitoring: an emerging nexus. Struct Health Monit 14(1):57–72
23. Sakagami T, Mizokami Y, Shiozawa D, Izumi Y, Moriyama A (2017) TSA based evaluation of fatigue crack propagation in steel bridge members. Procedia Struct Integrity 5:1370–1376

Chapter 14
Indentation Testing

14.1 Introduction

The indentation test is a simple commonly used technique to measure the hardness and related mechanical properties of materials in an easy and speedy way. The method consists of touching the material of interest with another material whose properties are known. In a typical test, a hard indenter of known geometry is driven into a soft material by applying a preset load and the dimensions of the resulting imprint are measured and related to the hardness index number.

The first systematic test to measure the hardness was proposed by mineralogist Mohs in 1822. It is based on the ability of one material to visibly scratch another. Materials that we're able to leave a permanent scratch on another were ranked harder with diamond assigned the maximum value of 10 on the scale. At the beginning of the twentieth century, indentation tests were performed by Brinell using spherical balls as indenters. Other indentation tests include the Knoop, Vickers, and Rockwell tests.

With the advent of nanotechnology, indentation was directed down to the nanometer range. This led to the development of nanoindentation which is a combination of high resolution recording indentation and the accompanying data analyses for the determination of mechanical properties directly from the load–displacement data without imaging the indentation. The principal goal of nanoindentation is to obtain the modulus of elasticity, hardness, and fracture toughness properties of thin films and small volumes of material. The forces involved are in the milli-Newton range and the depths of penetration are in the order of nanometers.

Indentation testing can be categorized into macro-indentation, micro-indentation, and nanoindentation depending on the magnitude range of the applied load. In the following, we present the principles of contact mechanics and the major methods used to measure the hardness, the modulus of elasticity, and the critical stress intensity factor.

14.2 Contact Mechanics

Consider an elastic sphere of radius R that intends an elastic half-plane under an applied load P (Fig. 14.1). The radius a of the circle of contact of the sphere and the half-plane is given (according to Hertz theory) by

$$a^3 = \frac{3}{4} \frac{P R}{E^*}$$

(14.1)

where

$$\frac{1}{E^*} = \frac{1 - v^2}{E} + \frac{1 - v'^2}{E'}$$

(14.2)

E and v are the modulus of elasticity and Poisson' ration for the half-plane and E' and v' are the corresponding quantities for the indenter. E^* combines the modulus of elasticity and Poisson's ratio of the half-plane and the indenter. It is often referred to as the "combined modulus" or "reduced modulus" of the system.

The total displacement of the indenter δ is given by

$$\delta = \frac{a^2}{R} = \left(\frac{9 P^2}{16 E^{*2} R} \right)^{1/3}$$

(14.3)

From Eq. (14.3) we obtain

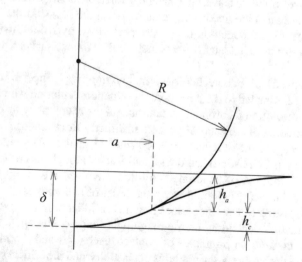

Fig. 14.1 Contact between a rigid indenter of radius R and a half-plane. The radius of the circle of contact is a and the total depth of penetration is δ. h_a is the depth of the circle of contact from the specimen free surface, and h_c is the distance from the bottom of the contact to the contact circle

$$P = \frac{4}{3} E^* R^{1/2} \delta^{3/2} \qquad (14.4)$$

Equation (14.4) expresses the load P versus displacement δ relation of the indenter. Note that P varies with δ in a power law with exponent 3/2.

14.3 Macro-indentation Testing

In macro-indentation tests the applied load P is in the range $2\,\text{N} < P < 30\,\text{kN}$. The major tests are the Brinell, Meyer, Vickers, and Rockwell tests. The tests determine the resistance of the material to penetration of a non-deformable indenter in the form of a ball, pyramid, or cone. We briefly present these tests.

14.3.1 Brinell Test

The material is indented by a hard spherical ball of diameter D through a fixed load P for a certain period of time (Fig. 14.2). The load is then removed and the chordal diameter d of the impression is measured with an optical microscope. The Brinell hardness number (BHN) is calculated as the load divided by the *actual area* A_c of the curved surface of the impression

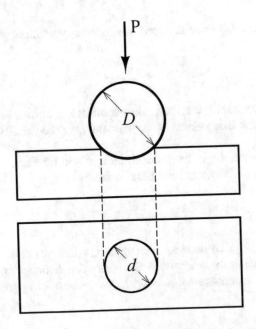

Fig. 14.2 Brinell indentation test

$$\mathrm{BHN} = \frac{P}{A_c} = \frac{2P}{\pi D\left(D - \sqrt{D^2 - d^2}\right)} \tag{14.5}$$

The load P is expressed in kilograms force. If Newton is used for the load the BHN must be divided by 9.81. The Brinell test has been standardized by the American Society for Testing and Materials (ASTM) and by the International Organization for Standardization (ISO). The load is applied for 10–30 s. The diameter d is taken as the mean value of two diameters of the impression at right angles. In a typical test, the diameter of the ball $d = 10$ mm and the applied load $P = 3000$ kgf (~29.4 kN). Smaller loads of $P = 1500$ kgf (~14.9 kN) and $P = 500$ kgf (~4.9 kN) are used for softer materials. Tests on small parts use balls of diameter of $D = 1$ mm and load of $P = 1$ kgf (~9.8 N).

14.3.2 Meyer Test

The Meyer test is based on the same principle as the Brinell test (Fig. 14.2). The test was originally defined for spherical indenters, but can be applied to any indenter shape. The Meyer hardness number (MHN) is expressed by the load divided by the *projected contact area* of the impression as

$$\mathrm{MHN} = \frac{P}{A_p} = \frac{4P}{\pi d^2} \tag{14.6}$$

From experiments, Meyer deduced the following relation which is known as Meyer's law

$$P = k\,d^n \tag{14.7}$$

where k is a proportionality constant. The exponent n is known as the Meyer index. It was found that it is independent of the diameter D of the ball. Its value is between 2 and 2.5.

The effective strain $\varepsilon_{\mathrm{eff}}$ of the indentation imposed by a spherical tip, as it was established experimentally, can be approximated by

$$\varepsilon_{\mathrm{eff}} = 0.2\frac{d}{D} \tag{14.8}$$

Equation (14.8) can be used to create the indentation stress–strain curve of a material by measuring the impression diameters for different applied loads. The yield stress σ_y for materials such as copper and steel can be approximated as $\sigma_y = \mathrm{MHN}/2.8$.

14.3.3 Vickers Test

The indenter in the Vickers test is a diamond in the form of a square-based pyramid. Its opposite sides meet at the apex at an angle of 136°, the edges at 148°, and the faces at 68° (Fig. 14.3). The Vickers hardness number (HV) is calculated from the ratio of the applied load P in kilograms force (kgf) and the actual surface area of the impression A_c in mm². A_c is given by

$$A_c = \frac{d^2}{2\sin\left(\frac{136°}{2}\right)} = \frac{d^2}{1.8544} \tag{14.9}$$

where d (mm) is the length of the diagonal measured from corner to corner on the residual impression of the specimen surface.

The Vickers hardness number is calculated by

$$HV = \frac{P}{A_c} = \frac{1.8544\,P}{d^2} \tag{14.10}$$

where P is measured in kgf and d in mm.

The applied loads vary in the range of 1–120 kgf with standard values of 5, 10, 20, 30, 50, 100, and 120 kgf. The time of application of the load is 10–15 s. The size of the impression is measured with a microscope with a tolerance of $\mp 1/1000$ mm. The Vickers contact area is related to the penetration depth t ($d = 7t$) by $A_c = 24.5t^2$ if

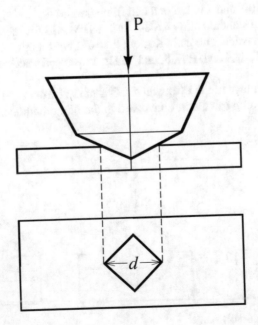

Fig. 14.3 Vickers indentation test

the elastic recovery of the material is not important. This allows calculation of VHN from measurement of the penetration depth t.

14.3.4   Rockwell Test

The previous three methods of indentation (Brinell, Meyer, Vickers) require measurement of the diameter of the indentation by an optical microscope. In the Rockwell test, the depth of penetration of an indenter under load into the sample is measured. The hardness is determined from the applied load on a spherical indenter and the depth of penetration. The test involves the application of a minor load of $P_0 = 10$ kgf (~98.1 N) followed by a major load P_1 (Fig. 14.4). The minor load establishes the zero position. It is used to eliminate errors in measuring the penetration depth. The major load is applied and removed, while the minor load is still maintained. The Rockwell hardness (HR) is calculated by

$$HR = N - 500h \qquad (14.11)$$

where N is a scale factor and h measured in mm is the difference of the penetration depths of the major and minor loads.

The value of N depends on the used indenter. It is 100 for spheroconical indenters and 130 for a ball. Equation (14.11) shows that the penetration depth and the hardness are inversely proportional.

Loads of 60, 100, and 150 kgf and ball diameters of ½, ¼, 1/8, 1/16 in. are used, as described in the standards ISO 6508-1 and ASTM E18 for metallic materials and ISO 2039-2 for plastics. The main Rockwell scales are established by letters as: A, B, C, D, E, F, G, H, K, L, M, P, R, S, and V. The most used scales are shown in Table 14.1.

The main advantage of the Rockwell test is that the hardness values are displayed directly, thus avoiding calculations involved in the other methods.

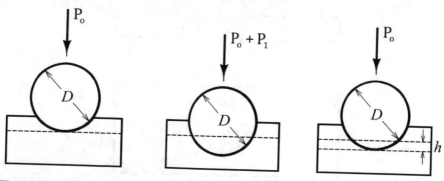

Fig. 14.4 Rockwell indentation test

Table 14.1 Main Rockwell scales

Scale	Name	Indenter	Load (kgf)
A	HRA	120° diamond spheroconical	60
B	HRB	1/16-in.-diameter (1.588 mm) steel sphere	100
C	HRC	120° diamond spheroconical	150
D	HRD	120° diamond spheroconical	100
E	HRE	1/8-in.-diameter (3.175 mm) steel sphere	100
F	HRF	1/16-in.-diamter (1.588 mm) steel sphere	60
G	HRG	1/16-in.-diameter (1.588 mm) steel sphere	150

14.4 Micro-indentation Testing

In micro-indentation tests, the applied load P is $P < 10$ N and the depth penetration h is $h > 0.2$ μm. The main tests are the Vickers and Knoop tests. As in the macro-indentation tests, the material resistance to the penetration of a diamond indenter with a shape of a pyramid under a given load and within a specific time period is determined.

14.4.1 Vickers Test

It is similar to the macro-indentation test, with the only difference that the applied load is smaller [<1 kgf (~9.81 N)]. The test is described by ISO 6507 and ASTM E384.

14.4.2 Knoop Test

The indenter is a rhombic-based pyramidal diamond that produces an elongated indent. The angles from the opposite faces of the indenter are 172.5° for the long edge and 130° for the short edge. The indenter produces a rhombic-shaped indentation with an approximate ratio of the long and short diagonals of 7–1.

The Knoop hardness number (KHN) is determined from the *projected area* A_p as

$$\text{KHN} = \frac{P}{A_p} = \frac{2P}{d^2\left(\cot\frac{172.5°}{2}\tan\frac{130°}{2}\right)} = 14.24\frac{P}{d^2} \tag{14.12}$$

where d is the length of the longest diagonal in mm and P is the indentation load in kgf.

The load is maintained for 10–15 s and then the indenter is removed leaving an elongated impression. A high-magnification microscope is used to measure the impression size. The applied load is in the range 10–1000 g (98 mN–9.8 N). A high-magnification microscope is needed to measure the indent size. The test is used particularly for very brittle materials or thin sheets.

14.5   Nanoindentation Testing

14.5.1   Introduction

Study of the mechanical properties of materials at the nanoscale range has received much attention in the last years due to the development of nanostructured materials and the application of nanometer thick films in engineering and electronic components. Nanoindentation is the combination of high resolution recording indentation and the accompanying data analyses for the determination of mechanical properties directly from the load–displacement data without imaging the indentation. A major difference between macro- and micro-indentation on one hand and nanoindentation on the other is that in nanoindentation the load and indentation depth is continuously monitored during the test. Nanoindentation employs ultra-low load indentation instruments (sensors and actuators) of high resolution capable of continuously monitoring the loads and displacements on an indenter as it is driven and withdrawn from a material. Loads of the order of a nano-Newton (10^{-9} N) and displacements of the order of an Angstrom (10^{-10} m) can be accurately measured.

Mechanical properties of materials in small dimensions can be very different from those of bulk materials having the same composition. From a nanoindentation test the elastic modulus, the hardness, and the fracture toughness of a material can be measured. These mechanical properties characterize the three fundamental modes of deformation of solids, elasticity, plasticity, and fracture, respectively.

In the following, we will first present a data analysis method for determining Young's modulus and hardness based on the elastic contact method. We will then present nanoindentation testing for measuring the fracture toughness of brittle materials at small volumes. For more information, the reader is referred to [1].

14.5.2   The Elastic Contact Method

A schematic representation, of a typical load, P, versus indenter displacement, h, data for an indentation experiment is shown in Fig. 14.5. For the analysis of the load–displacement curve we make the following assumptions [2]:

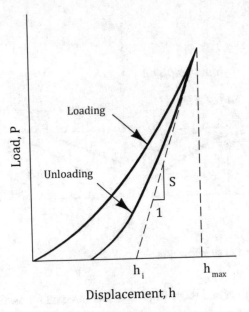

Fig. 14.5 Schematic representation of a typical load versus indenter displacement data for an indentation experiment

1. Deformation upon unloading is purely elastic.
2. The compliances of the specimen and the indenter tip can be combined as springs in series.
3. The contact can be modeled using the analytical model developed by Sneddon for the indentation of an elastic half space by a punch that can be described by an axisymmetric solid of revolution.

A cross section of an indentation is shown in Fig. 14.6. During loading, the total displacement h is written as

$$h = h_c + h_s \tag{14.13}$$

where h is the vertical distance along which contact is made (called contact depth), h_s is the vertical displacement of the surface at the perimeter of the contact and h_c is the penetration depth of the indenter under load. When the indenter is withdrawn the final depth of the residual hardness impression under load is h_f.

The determination of Young's modulus E^* defined in Eq. (14.2) is based on Hertz contact equation according to which

$$E^* = \frac{\sqrt{\pi}}{2} \frac{S}{\sqrt{A}} \tag{14.14}$$

where A is the contact area and S is the stiffness of the unloading curve ($S = dP/dh$) (Fig. 14.5).

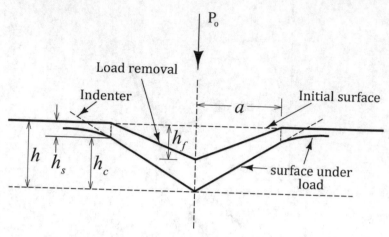

Fig. 14.6 Cross section of an indentation

According to Oliver and Pharr [2], the unloading data for stiffness measurement are fitted into equation

$$P = B(h - h_f)^m \qquad (14.15)$$

where P is the load, $(h - h_f)$ is the elastic displacement and B and m are material constants. The quantities B, m, and h_f are determined by the least squares fitting procedure of the unloading curve.

For the analysis, it is assumed that the geometry of the indenter is described by an area function $A = A(h)$ which relates the cross-sectional area of the indenter to the distance from its tip. The contact area at maximum load is given by

$$A = A(h_c) \qquad (14.16)$$

The contact depth at maximum load, h_c, that is, the depth along the indenter axis to which the indenter is in contact with the specimen is determined by

$$h_c = h_{max} - \varepsilon(h_{max} - h_i) \qquad (14.17)$$

where h_{max} is the maximum depth and h_i is the intercept depth, that is, the intercept of the tangent to the unloading load–displacement curve at maximum load with the depth axis. The constant ε is a function of the shape of the indenter tip. It takes the value *one* for a flat punch, the value of 0.7268 for a cone indenter, and the value of 0.75 for a spherical or paraboloidal indenter. The quantities h_{max} and h_i are determined from the load–displacement curve (Fig. 14.5).

The area function $A(h_c)$ depends on the shape of the indenter. For a Berkovich indenter (a three-sided pyramid with angles between the axis of symmetry and a face of 35.3°) it takes the form

$$A(h_c) = 24.54h_c^2 \qquad (14.18)$$

The hardness H_c is defined by

$$H_c = \frac{P_{max}}{A_c} \qquad (14.19)$$

where P_{max} is the peak indentation load and A_c is the contact area under maximum load.

This definition of hardness is different from that used in an imaging indentation test. In the latter case, the area is the residual area measured after the indenter is removed, while in the nanoindentation test the area is the contact area under maximum load. This distinction is important for materials with large elastic recovery, for example, rubber. A conventional hardness test with zero residual areas would give infinite hardness, while a nanoindentation test would give a finite hardness.

Equation (14.18) gives the area function $A(h_c)$ for an ideal Berkovich indenter. However, real tips are never ideally sharp and generally, are characterized by a radius of curvature at the tip. In such cases, the function $A(h_c)$ must be determined. Methods for determining $A(h_c)$ include the TEM (transmission electron microscope) replica method in which replicas of indentation are made and their areas are measured in TEM, the SFM (scanning force microscope) method in which the indenter tip is measured with a sharper SFM tip of known shape and a method based on Eq. (14.14) applied to a number of materials with known Young's modulus.

The above analysis suggests the following procedure for the determination of Young's modulus and hardness in nanoindentation from Eqs. (14.14) and (14.19):

1. Use Eq. (14.15) to fit the unloading data.
2. Find h_c from Eq. (14.17) using the value of depth at maximum load, h_{max}, the slope of the fit at P_{max} to obtain h_i, and the appropriate value of ε.
3. Use the area function $A(h_c)$ of the indenter to find the contact area at maximum load from the contact depth h_c.

From the value of E^* determined from Eq. (14.14) the elastic modulus, E, of the material is determined from Eq. (14.2) when the elastic modulus E' of the indenter is known.

14.5.3 Nanoindentation for Measuring Fracture Toughness

Nanoindentation testing may be applied to evaluate the fracture toughness of brittle materials at small volumes. The method was developed by Lawn et al. [3].

During elastic/plastic contact two types of cracks may form: those cracks which form on symmetry median planes containing the load axis and those which form on planes parallel to the specimen surface. The pertinent cracks used to determine fracture toughness are the first of these, the median/radial cracks. These cracks emanate from the edge of the contact impression, are oriented normal to the specimen surface on median planes coincident with the impression diagonals, and have a half-penny configuration (Fig. 14.7). It was observed that most of the crack development occurs not on loading, but on unloading the indenter. Thus, the main driving force for the formation of these cracks is the irreversible component of the contact stress. After unloading the indenter, characteristic radial traces are left on the specimen surface (Fig. 14.8). They provide the necessary information for the evaluation of fracture toughness.

The critical stress intensity factor, K_c, is given by Lawn et al. [3]

$$K_c = \alpha \left(\frac{E}{H} \right)^{1/2} \frac{P}{c^{3/2}}$$

(14.20)

where

E Young's modulus
H Hardness
P Peak indentation load
c Characteristic crack length (crack length plus half diagonal impression length)
α Empirical constant which depends on the geometry of the indenter ($\alpha = 0.016$ for a Vickers pyramidal indenter, $\alpha = 0.040$ for a cube-corner indenter).

Note that for the determination of K_{Ic} both E and H are needed. They can be determined from the analysis of the nanoindentation data. Thus, from one test all three quantities E, H and K_{Ic} can be determined in a straightforward way. For an

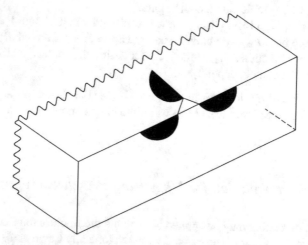

Fig. 14.7 Median/radial cracks emanating from the edge of the contact impression oriented normal to a specimen surface

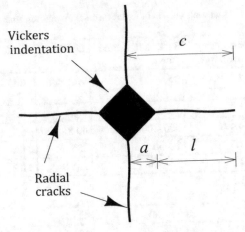

Fig. 14.8 Characteristic radial traces of the median/radial cracks left on the specimen surface after unloading

accurate determination of the crack length the test surface must be prepared to an optical finish. The method applies to those materials which produce a well-defined radial/median crack system.

In the application of the method, it should be mentioned that there are well-defined loads called cracking thresholds, which depend on the material and the type of indenter below which cracks do not develop. For most ceramic materials the cracking thresholds for Vickers and Berkovich indenters are about 250 mN or more. The crack lengths produced at these loads are relatively large which places limitations on the spatial resolution of the method. The cracking thresholds can be reduced by using indenters with smaller tip angles, for example, the cube-corner indenter has an angle of 35.3° between the axis of symmetry and a face, as compared to 65.3° for the Berkovich indenter. Using cube-corner indenters thresholds less than 10 mN can be achieved, while for Vickers indenters thresholds are of 1 N or greater.

Table 14.2 [4] shows values of Young's modulus, E, hardness, H, and fracture toughness, K_{Ic}, for various materials.

Table 14.2 Properties of materials used in indentation cracking measurement of fracture toughness (Reference 13.3)

Material	Cube-corner threshold (mN)	Vickers threshold (mN)	E^a (GPa)	H^a (GPa)	K_c (MPa$\sqrt{m}$)
Soda-lime glass	0.5–1.5	250–500	76.1	6.1	0.70[b]
Fused quartz	0.5–1.5	1000–1500	69.3	8.3	0.58[b]
Pyrex glass	1.5–4.4	500–1000	60.5	6.3	0.63[b]
Silicon (100)	0.5–1.5	20–50	185.6	11.5	0.7[c]
Silicon (111)	0.5–1.5	50–100	205.8	11.2	0.7[c]
Germanium (111)	0.5–1.5	<10	133.6	10.1	0.5[c]
Sapphire (111)	4.4–13.3	50–100	433.1	25.9	2.2[c]
Spinel (100)	4.4–13.3	100–150	286.2	18.4	1.2[c]
Silicon nitride (NC 132)	40–120	1000	319.9	21.6	4.7[c]
Silicon carbide (SA)	4.4–13.3	100–150	454.7	30.8	2.9[c]
Silicon carbide (ST)	Grain pushout	500–1000	427	21.8	4.1[c]

[a] Nanoindentation measurements with Berkovich indenter
[b] 3-pt bend chevron notch method
[c] From material data sheet or literature

Further Readings

1. Gdoutos EE (2020) Fracture mechanics, 3rd edn. Springer, Berlin, pp 371–385
2. Oliver WC, Pharr GM (1992) An improved technique for determining hardness an elastic modulus using load and displacement sensing indentation experiments. J Mat Res 7:1564–1583
3. Lawn BR, Evans AG, Marshall DB (1980) Elastic/plastic indentation damage in ceramics: the median/radial crack system. J Am Cer Soc 63:574–581
4. Pharr GM (1998) Measurement of mechanical properties of ultra-low load indentation. Mat Sci Eng A 253:151–159
5. Solomah AG (ed) (2004) Indentation techniques in ceramic materials characterization: theory and practice. Wiley, Hoboken
6. Bourhis EL, Morris DJ, Oyen ML, Schwaiger R, Staedler T (eds) (2008) Fundamentals of nanoindentation and nanotribology IV: volume 1049 (MRS proceedings). Cambridge University Press, Cambridge
7. Bahr DF, Morris DJ (2008) Nanoindentation: localized probes of mechanical behavior of materials. In: Sharpe WN Jr (ed) Handbook of experimental solid mechanics. Springer, Berlin, pp 389–407
8. Antunes J (2010) On depth sensing indentation of materials. VDM Verlag, Germany
9. Fischer-Cripps AC (2011) Nanoindentation, 3rd edn. Springer, Berlin
10. Alta K (2011) Effects of varying humiditiy in polymers by nanoindentation: investigation of the effects of varying humidity in additive manufactured by depth sensing indentation. LAP Lambert Academic Publishing
11. Navamathavan R, Nirmala R (2011) Mechanical properties of some III-V and II-VI semiconductor alloys: a micro and nanoindentation approaches. LAP Lambert Academic Publishing

283

12. Chen L (2015) Micro-nanoindentation in materials science. NY Research Press
13. Cagliero R (2016) Instrumented indentation test: in the macro hardness range. LAP Lambert Academic Publishing
14. Handadi UP, Udupa KR (2017) Indentation creep studies on stainless steel welds and solder alloys. Scholars' Press
15. Tiwari A, Natarajian S (2017) Applied nanoindentation of advanced materials. Wiley, Hoboken
16. Argatov I, Mishuris G (2018) Indentation testing of biological materials. Springer, Berlin
17. Tsui T, Pharr M (2018) Advanced nanoindentation of materials. Mdpi AG, Basel
18. Bhattacharyya A (2018) Substrate effect and nanoindentation failure. LAP Lambert Academic Publishing
19. Wang H, Zhu L, Xu B (2018) Residual stresses and nanoindentation testing of films and coatings. Springer, Berlin
20. Dey A, Mukhopadhyay AK (2018) Nanoindentation of natural materials: hierarchical and functionally graded microstructures. CRC Press, Boca Raton
21. Oyen ML (2019) Handbook of nanoindentation: with biological applications. Jenny Stanford Publishing
22. Tsui T, Volinsky A (2019) Small scale deformation using advanced nanoindentation techniques. Mdpi AG, Basel
23. Mohanty P, Behera A (2020) Nanoindentation study of NiTi thin film shape memory alloys: varying annealing temperature. LAP Lambert Academic Publishing
24. Murthy CSN (2021) Rock indentation: experiments and analyses. CRC Press, Boca Raton
25. Li X, Bhushan B (2002) A review of nanoindentation continuous stiffness measurement technique and its applications. Mat Charact 48:11–36
26. Oliver WC, Pharr GM (2004) Measurement of hardness and elastic modulus by instrumented indentation: advances in understanding and refinements to methodology. J Mater Res 19(1):3–20
27. Pathak S, Kalidindi SR (2015) Spherical nanoindentation stress-strain curves. Mat Sci Eng 91:1–36
28. Prasanna HU, Udupa KR (2016) Indentation creep studies to evaluate the mechanical properties of stainless steel welds. Aust J Mech Eng 14:39–43
29. Broitman E (2017) Indentation hardness measurements at macro-, micro-, and nanoscale: a critical review. Tribol Lett 65:1–18
30. Wen W, Becker AA, Sun W (2017) Determination of material properties of thin films and coatings using indentation tests: a review. J Mat Sci 52:12553–12373

Chapter 15
Nondestructive Testing (NDT)

15.1 Introduction

Nondestructive testing(NDT) refers to the science and technology of non-invasive methods of testing, evaluation, and characterization of materials, components, or systems without impairing their performance and serviceability. It provides techniques to detect and characterize flaws in materials and structures and plays an important role in the prevention of failure. The terms **nondestructive examination (NDE)**, **nondestructive inspection (NDI)**, and **nondestructive evaluation (NDE)** are also commonly used to describe this technology. NDT is important for the in-service inspection of load-bearing structures whose failure could have catastrophic consequences. In most NDT methods some form of energy, such as optical, electromagnetic, radiation, acoustic, etc. is sent through the material and the response of the material is analyzed by sensors. Sensor development played an important role and led to increased sensitivity and reliability of NDT methods.

The concept of fracture tolerance in fracture mechanics put new challenges to NDT methods. Structures are safe as long as the existing cracks do not surpass a critical size. In this respect, it became possible to accept structures containing defects under the condition that the sizes of the defects are smaller than a critical size. Detection of defects is not enough. The location, sizing, and orientation of the defects should be determined. Our ability to use fracture mechanics in design is largely due to the reliability of NDT methods. At the production or service inspection stage, parts containing flaws larger than those determined according to fracture mechanics design standards must be rejected or replaced.

In this chapter, we briefly present the following six major NDT methods: Dye penetrant, magnetic particles, eddy currents, X-ray diffraction, ultrasonics, and acoustic emission. Each of these methods possesses advantages and disadvantages depending on the application.

E. E. Gdoutos, *Experimental Mechanics*, Solid Mechanics and Its Applications 269, https://doi.org/10.1007/978-3-030-89466-5_15

15.2 Dye Penetrant Inspection (DPI)

15.2.1 Principle

Dye penetrant inspection involves application of a colored or fluorescent dye onto a cleaned surface of the component. After allowing sufficient time for penetration, the excess penetrant is washed off and the surface is dusted with a post-penetrant material (developer) such as chalk. The developer acts as a blotter and helps to draw penetrant out of the flaw. The defects show up under ultraviolet or white light, depending on the type of dye used (fluorescent or nonfluorescent (visible), respectively).

15.2.2 Application

Application of the method involves the following steps:

Pre-cleaning. The surface of the material to be tested is properly cleaned to remove any dirt, oil, etc. that could not allow the penetrant to go through the defect.

Application and removal of penetrant. Penetrants are classified as visible or fluorescent. Visible penetrants use a color contrast (usually red) dye, while fluorescent penetrants use a dye that fluoresces under dark light. Fluorescent penetrants are more sensitive than visible dye penetrants. The penetrant is applied to the surface of the body under study for a "dwell time" (5–30 min) to soak into the flaws. The dwell time depends on the type of the penetrant, the material tested, and the size of the flaw (smaller flaws require a longer time). The excess penetrant is removed from the surface.

Application of developer. After the removal of the excess penetrant, a developer is applied. The function of the developer is to draw penetrant out of the defect onto the surface. A visible indication known as bleed-out is formed on the surface and indicates the location, orientation, and possible types of defects.

Inspection. Visible light is used for inspection for visible dye penetrant and ultraviolet radiation for fluorescent penetrants. The inspection time depends on the penetrant and developer used.

15.2.3 Advantages and Disadvantages

Dye penetrant is one of the simplest, but a highly sensitive method for the detection of defects open to the surface. It is well-suited for detection of all types of defects. The reliability of the method depends on the surface preparation of the component. A main advantage of the method is the speed of the test and the low cost. The method is applied to almost all materials, except porous materials, and can detect small cracks. Proper cleaning of the surface is required. The main disadvantage of the method is that it applies to only surface cracks and it is difficult to apply on rough and porous surfaces. The method is widely used.

15.3 Magnetic Particles Inspection (MPI)

15.3.1 Principle

The method of MPI is based on the principle that flaws in a magnetic material produce a distortion to an induced magnetic field. The body under examination can be magnetized directly when an electric current is passed through it, or indirectly when a magnetic field is applied from an external source. The lines of magnetic flux are perpendicular to the direction of the electric current, which may be alternating or direct. The presence of flaws causes the magnetic flux to leak since air cannot support a magnetic field, as metals do. Measuring this distortion provides information on the existing defects. For detecting the distortion of the magnetic field the surface under inspection is coated with a fluorescent liquid that contains magnetic particles in suspension. After the inspection the part is demagnetized. This is done with equipment that works the opposite way of the magnetizing equipment.

15.3.2 Advantages and Disadvantages

MPI has the added advantage over dye penetrant inspection in detecting subsurface defects. The method applies only to ferromagnetic materials, such as iron, nickel, cobalt, and some of their alloys. The inspection should be carried out with a magnetic field perpendicular to the plane of the flaw. More than one evaluation is needed to cover all orientations of the flaws. For thick parts, large currents are required. MPI is easy to apply, speedy and economical.

15.4 Eddy Currents Inspection (ECI)

15.4.1 Principle

Eddy currents are loops of electric current induced in conductors by a moving or varying magnetic field in the conductor. Eddy currents flow in closed loops within conductors in planes perpendicular to the magnetic field. Because their flow patterns resemble swirling eddies or whirlpools in a river, they are called eddy currents.

A coil of wire carrying alternating current is placed near a conducting surface. A magnetic field is produced. When the field interacts with a conductive object, eddy currents are induced in the material. Changes in the electrical conductivity and magnetic permeability of the test object because of defects result in a change in the eddy current. These changes are detected by impedance changes in the coil. By measuring this change we can find information about the defect. The induced eddy currents concentrate near the surface of the conductor; this is the so-called "skin effect". The penetration depth is influenced by the frequency of the current, the magnetic permeability and electrical conductivity of the conductor, and the coil and conductor geometry.

The sensitivity of the method is high for defects near the surface but decreases with increasing depth. Problems in the method arise from the difficulty of relating the defect size to the change in impedance, and the influence of a number of other factors on the impedance. These include: the relative position of the coil and the conductor; the presence of structural variations; material inhomogeneities. Measurement of defect size is made by comparing its effect to that observed from a standard defect.

15.4.2   Advantages and Disadvantages

ECI can detect surface defects with good accuracy. It is safe and does not produce any harmful radiation. Little pre-cleaning or post-cleaning is required. The process can be automated. ECI is susceptible to magnetic permeability changes, can be applied only to conductive materials, and is not suitable for large areas and complex geometries.

15.5   X-ray Diffraction (XRD)

15.5.1   Introduction

X-ray diffraction (XRD) is a nondestructive testing method for measuring strains and detecting subsurface defects. When X-rays are pointed at an object at an incident angle, they slightly penetrate the material and interact with the atoms. The scattered rays provide a pattern of the atoms inside the material. From the analysis of the intensity of the scattered pattern, the atomic and molecular structure of materials can be determined. Defects absorb less X-rays than the surrounding material, and therefore, they can be detected. In the following, we will briefly present the nature of X-rays, their diffraction with atoms/molecules inside a material, and their use in measuring strains.

15.5.2 X-rays

X-rays are high-energy electromagnetic radiation of very short wavelength capable to pass through many materials opaque to light. Most X-rays have wavelength ranging from 10×10^{-12} to 10×10^{-9} m and corresponding frequencies ranging from 30×10^{18} to 30×10^{15} Hz. In the electromagnetic range (Fig. 2.15) X-rays have wavelengths shorter than those of the UV rays and longer than those of the gamma rays. X-rays were discovered by W. C. Roentgen in 1895. The name of X-rays comes from their unknown character at the time of their discovery after the symbol x used in algebra to denote an unknown quantity. X-rays are more energetic and can be used to probe the inter-atomic distance in crystalline materials.

   X-rays are generated in an X-ray vacuum tube. An electrical current is used to heat a tungsten filament inside the tube. An electron beam is created by a high voltage and is allowed to strike a target (glass or metal) surface inside the tube. When this electron beam collides with the target surface X-rays are produced. Since their wavelength is very small they show no interference or diffraction effects with ordinary gratings of pitch of 10^{-6} m (10^3 nm).

15.5.3 X-ray Diffraction

It was proposed by Max von Laue in 1912 that a crystalline solid in which the atoms are arranged in regular patterns with spacing between neighboring atoms of the order of 10^{-10} m might serve as a three-dimensional diffraction grating for X-rays. A beam of X-rays is scattered, that is, absorbed and re-emitted by the atoms of the crystalline body. The scattered waves might interfere in the same way of a diffraction grating studied in Sect. 2.9.4. X-rays offer greater resolution than visible light because they have shorter wavelengths. They are very effective in studying the microscopic world of atoms and molecules. X-ray diffraction can be used to study the structure of crystals and for measuring strains.

15.5.4 Measurement of Strain

Consider a crystal in which the atoms are arranged in a cubical fashion. Let the distance between neighboring atoms is d (Fig. 15.1). A beam of X-rays in incident on the crystal at an angle φ with the surface. The two rays shown in Fig. 15.1 are reflected from two subsequent planes of atoms at a distance d apart. The optical path length difference between the two rays is $2d\sin\varphi$. Constructive interference occurs when $2d\sin\varphi$ is a multiple of wavelengths λ, that is

$$2d \sin \varphi = m\lambda \quad (m = 0, 1, 2, \ldots) \tag{15.1}$$

Fig. 15.1 X-ray diffraction by a crystal for strain measuring

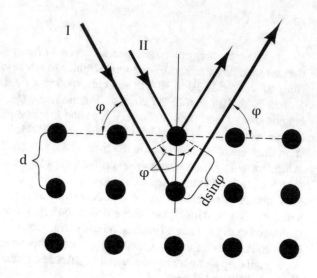

Equation (15.1) is called the **Bragg equation**. It can be used for the determination of the distance d between adjacent atoms when the angle of incidence φ and the wavelength λ are known. Equation (15.1) can also be used for a white or polychromatic X-ray beam and the wavelengths scattered at a particular direction are measured. *Equation (15.1) is the basis for strain measurement by X-ray diffraction.*

From Eq. (15.1) the strain ε is calculated as

$$\varepsilon = \frac{d - d_0}{d_0} = \frac{\sin \varphi_0}{\sin \varphi} - 1 \qquad (15.2)$$

where d and φ are the distance between two adjacent atoms and the angle of incidence with respect to the surface of the crystal of the stressed body, and d_0 and φ_0 are the corresponding quantities for the unstressed body. The spacing d between two successive planes in the crystal is measured along the direction that bisects the incident and diffracted X-ray beams.

Equation (15.2) refers to strain at a given direction. By using six different directions defined by the angle of incidence φ we can determine the corresponding normal strains. From the equations of strain transformation, we can then determine the six components of the strain tensor. From the strains, the stresses can be determined by using the appropriate constitutive equations.

15.5.5 Instrumentation

X-rays are generated when electrons accelerated by a high voltage in a vacuum tube strike a metal surface inside the tube. The vacuum tube consists of a cathode and an anode of a high melting heavy metal. Electrons are emitted from the cathode and accelerated to the anode. Voltages in the order of 30–150 kV are typically used. X-rays have wave-particle duality. The energy E of photons of X-rays and the frequency f of their wave nature are related by the equation

$$E = hf = h\frac{c}{\lambda} \tag{15.3}$$

where h is Plank's constant ($h = 6.626 \times 10^{-34}$ J s), c is the speed ($c = 3 \times 10^8$ m/s), f is the frequency and λ is the wavelength.

Introducing the values of c and h, Eq. (15.3) can be put in the form

$$E = \frac{12.4}{\lambda} \tag{15.4}$$

where λ is in Angstroms and E is in keV.

In applications, X-rays are produced by diffractometers which are laboratory-based or portable. A diffractometer is an X-ray stress analysis equipment consisting of an X-ray source and the capability for measuring diffraction angles. Synchrotron facilities provide high-energy X-rays with fluxes orders of magnitude higher than X-ray tubes and offer excellent capabilities for strain measurements. They can penetrate depths over 1 mm in most materials.

X-ray strain measurements based on laboratory equipment are restricted to small penetration depths typically a few tens of microns into materials yielding strain near the surface of the body. For deeper depths, destructive layer removal or more high-energy X-rays are used. X-rays are widely used for the determination of residual stresses.

15.6 Ultrasonic Testing (UT)

15.6.1 Introduction

Ultrasonic Testing (UT) uses high frequency sound waves well above the range of human hearing that travels in the object to be tested to detect internal flaws or to characterize materials. In most applications, the frequencies of the waves range from 0.1 to 25 MHz. UT can be used to detect flaws and discontinuities on steel, other metals, and alloys, as well as on concrete, composites, and wood, however, with less resolution. It is based on the interaction of ultrasonic waves and flaws. UT is one of the most widely used methods of nondestructive testing. It has been used for inspecting welds.

Ultrasound is used in medicine to create images of soft tissue structures, such as the gall bladder, liver, heart, kidney, female reproductive organs, and even of babies still in the womb. It cannot be used to image bones, because they are too dense to penetrate.

15.6.2 Operation

A UT inspection system consists of a pulser/receiver, a transducer, and display devices. A pulser is an electronic device that produces high voltage electrical pulses. The transducer is driven by the pulser to generate ultrasonic waves which propagate through the material. The ultrasonic waves follow the laws of optic waves in reflection, refraction, etc. During their travel in the material, they are reflected at interfaces. They are picked up by the receiver which transforms the wave signal into an electrical signal. The analysis of this signal provides information about the flaws and microstructures of the object under inspection.

There are two methods of receiving the ultrasound waves: reflection (or pulse-echo) and attenuation (or through transmission). In the *reflection method,* the sound is reflected back to the device from an interface, such as the back wall of the object or from an imperfection. The transducer (pulser and receiver) sends and receives the waves. The reflected wave signal is transformed into an electrical signal by the transducer and is displayed on a screen. Signal travel time can be directly related to the distance that the signal traveled. From the signal, information about the reflector location, size, orientation, and other features can be gained. In the *attenuation method,* the pulser sends the ultrasound and a separate receiver detects the amount that has reached it. When discontinuities exist between the transmitter and the receiver the amount of transmitted sound is reduced. From this amount, the presence of discontinuities is revealed.

15.6.3 Advantages and Disadvantages

Advantages of UT include: sensitivity of the method to both surface and subsurface discontinuities, higher depth of penetration compared to other NDT methods, need for one-sided access, detection of material thickness, accuracy in estimating position, size, and shape of discontinuity, minimal part preparation, instantaneous results using electronic equipment. A great advantage of UT is that it is a safe method. It does not pose a health hazard to technician, as opposed to radiographic testing which uses high-energy X-rays.

Limitations of UT include: accessibility of surface to transmit ultrasound and requirement of a coupling medium to promote the transfer of sound energy into the material. Furthermore, linear defects parallel to the sound beam may go undetected. Very small, rough, irregular in shape, very thin, or coarse grain materials, like cast iron, are difficult to inspect due to low sound transmission and high signal noise. A skilled and trained operator is needed.

15.7 Acoustic Emission Testing (AET)

15.7.1 Introduction

When a solid body is subjected to an external stimulus (loads, change of temperature, etc.) irreversible changes in its internal structure may occur resulting in sudden release of elastic energy in the form of stress waves. A kind of sound is usually produced. In some cases the sounds are loud. In other cases, the sounds are more subtle and for their detection specialized equipment is needed. The stress waves propagate in the body and result in small transient surface displacements which are recorded by sensors to provide information about the damage in the body. Events that generate AE include: dislocations, cracks, phase transformation, twinning, thermal stresses, cracking during cooldown, matrix cracking, fiber breakage, and debonding in composites.

Acoustic emission testing (AET) is a nondestructive method for the detection, recording, and analysis of acoustic emission (AE) signals using specialized equipment with the objective to obtain valuable information regarding the damage in the body. AET is of great importance in structural health monitoring (to detect, locate and characterize damage), quality control, process monitoring, and other fields. AET has been used as early as 6500 BC by potters to listen for audible sounds during the cooling of their ceramics, signifying structural failure.

Sources of AE vary from natural events like earthquakes to the initiation and growth of cracks, dislocation movements, cavitation processes, and phase transformations in metals. In composite materials matrix cracking, fiber breakage, and debonding contribute to AEs. Acoustic signals have also been recorded in polymers, wood, and concrete, among other materials.

AET differs from other nondestructive testing methods in two regards: First, energy is not supplied to the test object. AET listens to the energy released by the body while in operation. Second, AET deals with dynamic processes in the body. Only active damages (e.g., crack growth) are detected. Defects go undetected if they cannot create an acoustic event.

15.7.2 Acoustic Emission Testing

AE elastic waves propagate in all directions and hit a piezoelectric sensor that is attached to the surface of the body. They contain information about the internal behavior of the material and the geometry of the body. The wave motion at the surface of the body on picometer to nanometer scale is converted into an electric signal by the piezo element. Signals can be detected at frequencies under 1 kHz. Typically, most of the released energy is in the range of 1 kHz to 1 MHz. The sensitivity of the sensor is key issue for detection of AE signals and source location.

For testing small components only one sensor may be sufficient. Typically, multiple sensors are used. Different sensors pick up different signal characteristics for the same acoustic emission event. A pattern of interlocking triangles or rectangles is used to set up sensors. A fluid couplant is used to bond the sensor to the surface and help the sensor to obtain stronger signal. The sensors are connected to an amplifier and additional electronic equipment is used to filter and isolate the sound.

The AET system records any AE signals along with the exact time it occurred. Data related to emission count and length, signal length, peak amplitude, and other parameters are recorded. After the test is completed the results are analyzed. By measuring the arrival time of an AE signal to each sensor the location of the defect can be determined by knowing the velocity of the wave in the material and the difference in arrival times among the sensors. Following determination of the location, additional testing can be performed by other nondestructive methods for measuring the size and inclination of the defect.

15.7.3   Advantages and Disadvantages

AET can be used for early detection and real-time monitoring of defects. Major advantages of the method include: early detection of small defects, no need to shut down the unit that can be inspected while in operation, detection of only active defects which may impose an immediate threat to the structure, reduced cost.

AET can provide only qualitative, not quantitative, results. It can detect the existence of a defect, but cannot determine its size and orientation. This requires other test methods. It can identify only active defects which can be an advantage, but in some cases, identification of stagnant defects is needed. Loud environments present challenges in the accuracy of the results and filtering of the noise is required. Finally, AET requires an experienced and skilled operator.

Further Readings

1. Summerscales J (1990) Non-destructive testing of fibre-reinforced plastic composites. Elsevier, Amsterdam
2. Halmshaw R (1991) Non-destructive testing, 2nd edn. Butterworth-Heinemann, Oxford
3. Doherty JE (1993) Nondestructive evaluation. In: Kobayashi AS (ed) Handbook of experimental mechanics, 2nd edn. Society for Experimental Mechanics, Bethel, pp 527–555
4. Blitz J, Simpson G (1995) Ultrasonic methods for non-destructive testing. Springer, Berlin
5. Raj B, Subramanian CV, Jayakumar T (2000) Non-destructive testing of welds. Woodhead Publishing, UK
6. Mix PE (2005) Introduction to nondestructive testing: a training guide. Wiley, Hoboken
7. Rao BPC (2006) Practical non-destructive testing. Alpha Science Int'l Ltd, UK
8. Raj B, Jayakumar T, Thavasimuthu T (2007) Practical non-destructive testing, 3rd edn. Alpha Science, UK

9. Almer JD, Winholtz RA (2008) X-ray stress analysis. In: Sharpe WN Jr (ed) Handbook of experimental solid mechanics. Springer, Berlin, pp 801–820

10. Prasad J, Nair CCK (2011) Non-destructive testing and evaluation of materials, 2nd edn. McGraw-Hill, New York

11. Zoughi R (2012) Microwave non-destructive testing and evaluation principles. Springer, Berlin

12. Blitz J (2012) Electrical and magnetic methods of non-destructive testing, 2nd edn. Springer, Berlin

13. Halmshaw R (2012) Industrial radiology: theory and practice, 2nd edn. Springer, Berlin

14. Czichos H (ed) (2013) Handbook of technical diagnostics. Springer, Berlin

15. Wong BS (2014) Non-destructive testing—theory, practice and industrial applications. LAP Lambert Academic Publishing

16. Huang S, Wang S (2016) New technologies in electromagnetic non-destructive testing. Springer, Berlin

17. Kleinert W (2016) Defect sizing using non-destructive ultrasonic testing. Springer, Berlin

18. Balayssac J-P, Garnier V (eds) (2017) Non-destructive testing and evaluation of civil engineering structures. ISTE Press-Elsevier

19. Murashov V (2017) Non-destructive testing and evaluation designs by the acoustic methods: quality-reliability-safety. LAP Lambert Academic Publishing

20. Ida N, Meyendorf N (eds) (2018) Handbook of advanced non-destructive evaluation. Springer, Berlin

21. Barkanov EN, Dumitrescu A, Parinov IA (eds) (2018) Non-destructive testing and repair of pipelines. Springer, Berlin

22. Bowler N (2019) Eddy-current nondestructive evaluation. Springer, Berlin

23. Papaelias M, Garcia Marquez FP, Karyotakis A (2019) Non-destructive testing and condition monitoring techniques for renewable energy industrial assets. Elsevier, Amsterdam

24. Ohtsu M (2020) Acoustic emission and related non-destructive evaluation techniques in the fracture mechanics of concrete: fundamentals and applications. Woodhead Publishing, UK

25. Cavalcanti WL, Brune K, Noeske M, Tserpes K, Ostachowicz W, Schlag M (eds) (2021) Adhesive bonding of aircraft composite structures: non destructive testing and quality assurance concepts. Springer, Berlin

26. Breysse D, Balayssac J-P (2021) Non-destructive in situ strength assessment of concrete. Springer, Berlin

Chapter 16
Residual Stresses—The Hole-Drilling Method

16.1 Introduction

Residual stresses are locked-in, self-equilibrating stresses that remain in a material after the external loads are removed. They result from any mechanisms that cause misfits among different parts of a material or structure, such as processing operations, non-uniform plastic deformation, temperature gradients, surface treatments, material forming, and shaping procedures, phase transformations, etc. Residual stresses are developed in composite materials, welds, quenched components, semiconductor fabrication, thin films, etc. They can be tensile or compressive. They are algebraically summed with applied stresses. In some cases, beneficial compressive residual stresses are introduced intentionally, as in pre-stressed concrete, in brittle materials which can be toughened, in shot peening, quenching, tempered glass. Generally speaking, residual stresses are undesirable.

Because of their hidden nature, residual stresses are difficult to measure. Measurement methods can be categorized as destructive and nondestructive. The destructive methods involve measuring the deformations by cutting off some parts of the component. The damage occurred is usually minimal, even though in some cases may be quite extensive. Among the destructive methods, the most popular is the **hole-drilling method** which is the subject of this chapter. In this method, a small hole is drilled in the material typically 1–4 mm in diameter. The method may be considered semi-destructive because the diameter of the hole is small. Nondestructive methods include X-ray, neutron diffraction, ultrasonic and magnetic methods. These methods have briefly been developed in Chap. 15 and will not be repeated here.

In the following, the hole-drilling method of measuring residual stresses will be presented. The method involves monitoring the change of the relieved strains produced when a hole is drilled in a component containing residual stresses. We will present the method and the relevant equations for the determination of the residual stresses. The cases of uniaxial and biaxial residual stresses will be studied separately.

© The Author(s), under exclusive license to Springer Nature Switzerland AG 2022
E. E. Gdoutos, *Experimental Mechanics*, Solid Mechanics and Its Applications 269,
https://doi.org/10.1007/978-3-030-89466-5_16

16.2   Hole-Drilling Method

The hole-drilling method using strain gages for the measurement of the relieved strains involves the following steps:

1. Installation of the strain gage rosette.
2. Drilling of a hole at the center of the rosette in a series of predetermined depth increments.
3. Monitoring the strain gage readings after each depth increment.
4. Using proper equations for the calculation of the residual stresses from the recordings of the strain gages.

16.3   Uniaxial Residual Stresses

We will first analyze the simple case of a uniaxial residual stress field in a thin plate. A through-the-thickness hole is drilled in the plate and the relieved strains on the surface of the plate are recorded with strain gages. We will relate the residual stress with the recorded strains.

For this reason, consider a thin infinite plate subjected to a uniform uniaxial residual stress σ_0 along the y-axis (Fig. 16.1). The normal polar stresses $\sigma_{rr}^I, \sigma_{\theta\theta}^I$ at point $P(r, \theta)$ of the plate are given by

$$\sigma_{rr}^I = \frac{\sigma_0}{2}(1 - \cos 2\theta)$$

$$\sigma_{\theta\theta}^I = \frac{\sigma_0}{2}(1 + \cos 2\theta) \tag{16.1}$$

A through-the-thickness hole of radius is drilled in the plate. The stress distribution after drilling the hole is obtained from the Kirsch solution of an infinite plate with a hole subjected to uniaxial stress. We have for the normal polar stresses $\sigma_{rr}^{II}, \sigma_{\theta\theta}^{II}$ at point $P(r,\theta)$ of the plate

$$\sigma_{rr}^{II} = \frac{\sigma_0}{2}\left\{\left(1 - \frac{a^2}{r^2}\right)\left[1 + \left(\frac{3a^2}{r^2} - 1\right)\cos 2\theta\right]\right\}$$

$$\sigma_{\theta\theta}^{II} = \frac{\sigma_0}{2}\left[\left(1 + \frac{a^2}{r^2}\right) + \left(1 + \frac{3a^4}{r^4}\right)\cos 2\theta\right] \tag{16.2}$$

where $r > a$.

The change in normal stresses, $\sigma_{rr}, \sigma_{\theta\theta}$, produced when the hole is drilled is obtained by subtracting the initial stresses $\sigma_{rr}^I, \sigma_{\theta\theta}^I$ given by Eq. (16.1) from the stresses $\sigma_{rr}^{II}, \sigma_{\theta\theta}^{II}$ after drilling the hole given by Eq. (16.2) as

$$\sigma_{rr} = \sigma_{rr}^{II} - \sigma_{rr}^I = -\frac{\sigma_0 a^2}{2r^2}\left[1 + \left(\frac{3a^2}{r^2} - 4\right)\cos 2\theta\right]$$

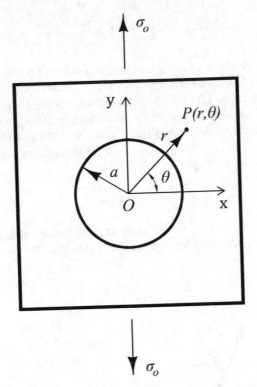

Fig. 16.1 An infinite plate subjected to a uniform uniaxial residual stress σ_0 along the y-axis

$$\sigma_{\theta\theta} = \sigma_{\theta\theta}^{II} - \sigma_{\theta\theta}^{I} = \frac{\sigma_0 a^2}{2r^2}\left(1 + \frac{3a^2}{r^2}\cos 2\theta\right) \tag{16.3}$$

The change in normal strains, ε_{rr}, $\varepsilon_{\theta\theta}$, produced when the hole is drilled in the plate is obtained from the stresses σ_{rr}, $\sigma_{\theta\theta}$ using Hooke's law and assuming conditions of plane stress (the thickness of the plate is small and the hole is drilled through the plate thickness). We obtain

$$\varepsilon_{rr} = \frac{1}{E}(\sigma_{rr} - \nu\sigma_{\theta\theta}) = -\frac{\sigma_0 a^2(1+\nu)}{2Er^2}\left(1 + \frac{3a^2}{r^2}\cos 2\theta - \frac{4\cos 2\theta}{1+\nu}\right)$$
$$\varepsilon_{\theta\theta} = \frac{1}{E}(\sigma_{\theta\theta} - \nu\sigma_{rr}) = -\frac{\sigma_0 a^2(1+\nu)}{2Er^2}\left(1 + \frac{3a^2}{r^2}\cos 2\theta - \frac{4\nu\cos 2\theta}{1+\nu}\right) \tag{16.4}$$

Equation (16.4) for the strains $\varepsilon_{rr} = \varepsilon_{xx}$, $\varepsilon_{\theta\theta} = \varepsilon_{yy}$ along the x-axis ($\theta = 0, r = x$) takes the form

$$\frac{2E\varepsilon_{xx}}{\sigma_0(1+\nu)} = -\left(\frac{a}{r}\right)^2\left(1 + \frac{3a^2}{r^2} - \frac{4}{1+\nu}\right)$$

$$\frac{2E\varepsilon_{yy}}{\sigma_0(1+\nu)} = -\left(\frac{a}{r}\right)^2\left(1 + \frac{3a^2}{r^2} - \frac{4\nu}{1+\nu}\right) \tag{16.5}$$

Using Eq. (16.5) with $\nu = 1/3$ the normalized strains $1.5\, E\varepsilon/2\sigma_0$ ($\varepsilon = \varepsilon_{xx}, \varepsilon_{yy}$) are plotted in Fig. 16.2. Note the steep variation of strains along the x-axis for $1 < x/a < 1.5$. However, at $x/a = 1.75$ the strain ε_{xx} does not present a substantial variation with x, and therefore when a gage is placed at that point along the x-axis the effect of the gage length on measuring the strain is insignificant and can be omitted. We obtain from the first Eq. (16.5) for $x/a = 1.75$ for the residual stress σ_0 for a strain $\varepsilon_g = \varepsilon_{xx}$ measured by the gage

$$\sigma_0 = 4.5\, E\varepsilon_g \tag{16.6}$$

Equation (16.6) can be used for the calculation of the residual stress σ_0 when the strain ε_g at the point $x/a = 1.75$ along the x-axis is recorded.

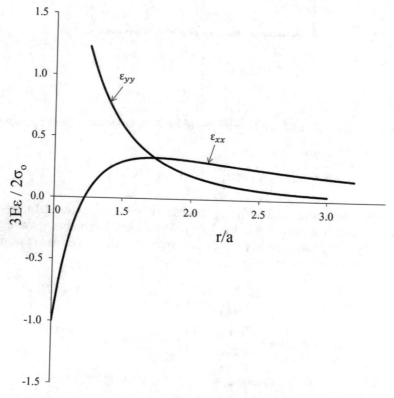

Fig. 16.2 Normalized strains $1.5E\varepsilon/2\sigma_0$ along the x-axis due to residual stress relief by hole drilling

16.4   Biaxial Residual Stresses

The previous analysis of the relation between the residual stresses and the relieved strains on the surface of the body concerned with the simple case of a uniaxial residual stress field and the length of the gage was ignored. In the present analysis, we will consider a biaxial residual stress field and will take into consideration the length of the strain gage.

Consider a biaxial residual stress field in a plate dictated by the values of the principal stresses σ_1 and σ_2 along the x- and y-axis, respectively (Fig. 16.3). We will relate the residual stresses with the change of strains on the surface of the plate following the same procedure as in the previous case of uniaxial residual stresses. The normal polar stresses σ_{rr}^I, $\sigma_{\theta\theta}^I$ at a point $P(r, \theta)$ of the plate are given by

$$\sigma_{rr}^I = \frac{\sigma_1 + \sigma_2}{2} + \frac{\sigma_1 - \sigma_2}{2} \cos 2\theta$$

$$\sigma_{\theta\theta}^I = \frac{\sigma_1 + \sigma_2}{2} - \frac{\sigma_1 - \sigma_2}{2} \cos 2\theta \tag{16.7}$$

A through-the-thickness hole of radius is drilled in the plate. The stress distribution after drilling the hole is obtained from the Kirsch solution of an infinite plate with a hole subjected to a biaxial stress field. We have for the polar stresses σ_{rr}^{II}, $\sigma_{\theta\theta}^{II}$ at point $P(r, \theta)$ of the plate

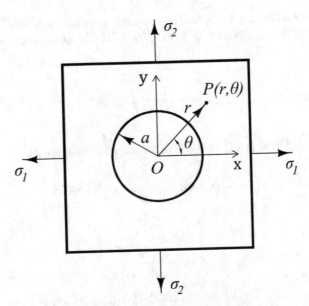

Fig. 16.3 An infinite plate subjected to uniform biaxial residual stresses σ_1 and σ_2 along the x- and y-axis, respectively

$$\sigma_{rr}^{II} = \frac{\sigma_1 + \sigma_2}{2}\left(1 - \frac{a^2}{r^2}\right) + \frac{\sigma_1 - \sigma_2}{2}\left(1 + \frac{3a^4}{r^4} - \frac{4a^2}{r^2}\right)\cos 2\theta$$

$$\sigma_{\theta\theta}^{II} = \frac{\sigma_1 + \sigma_2}{2}\left(1 + \frac{a^2}{r^2}\right) - \frac{\sigma_1 - \sigma_2}{2}\left(1 + \frac{3a^4}{r^4}\right)\cos 2\theta \qquad (16.8)$$

where $r > a$.

The change in normal stresses, $\sigma_{rr}, \sigma_{\theta\theta}$, produced when the hole is drilled is obtained by subtracting the initial stresses $\sigma_{rr}^{I}, \sigma_{\theta\theta}^{I}$ given by Eq. (16.7) from the stresses $\sigma_{rr}^{II}, \sigma_{\theta\theta}^{II}$ after drilling the hole given by Eq. (16.8) as

$$\sigma_{rr} = -\frac{\sigma_1 + \sigma_2}{2}\frac{a^2}{r^2} + \frac{\sigma_1 - \sigma_2}{2}\left(\frac{3a^4}{r^4} - \frac{4a^2}{r^2}\right)\cos 2\theta$$

$$\sigma_{\theta\theta} = \frac{\sigma_1 + \sigma_2}{2}\frac{a^2}{r^2} - \frac{\sigma_1 - \sigma_2}{2}\frac{3a^4}{r^4}\cos 2\theta \qquad (16.9)$$

The change in radial strain ε_{rr} produced when the hole is drilled in the plate is obtained from stresses $\sigma_{rr}, \sigma_{\theta\theta}$ using Hooke's law and assuming conditions of plane stress (the thickness of the plate is small and the hole is drilled through the plate thickness). We obtain

$$\varepsilon_{rr} = \frac{1}{E}(\sigma_{rr} - \nu\sigma_{\theta\theta}) \qquad (16.10)$$

The average radial strain recorded by the gage when the strain gradient along the length of the strain gage extending from r_1 to r_2 is taken into consideration is given by

$$\varepsilon_{rr} = \frac{1}{r_2 - r_1}\int_{r_1}^{r_2} \varepsilon_{rr}\, dr \qquad (16.11)$$

From Eqs. (16.9) to (16.11) we obtain for the relieved strain ε_{rr}

$$\varepsilon_{rr} = \frac{A}{E}(\sigma_1 + \sigma_2) + \frac{B}{E}(\sigma_1 - \sigma_2)\cos 2\theta \qquad (16.12)$$

where

$$A = -\frac{1 + \nu}{2}\frac{a^2}{r_1 r_2} \qquad (16.13)$$

$$B = \frac{2a^2}{r_1 r_2}\left[-1 + \frac{1 + \nu}{4}\frac{a^2(r_1^2 + r_1 r_2 + r_2^2)}{r_1^2 r_2^2}\right] \qquad (16.14)$$

For the determination of the residual stresses the magnitudes of the stresses σ_1 and σ_2 and their angle of inclination need to be determined. The three unknown quantities are calculated from the readings of three radial strain gages placed on the surface of the body near the hole. For the case of a three-element strain gage rosette whose elements are at angles θ, $(\theta + 45°)$ and $(\theta - 45°)$ (Figs. 16.4 and 16.5), we obtain from Eq. (16.12)

$$\varepsilon_0 = K_1\varepsilon_0' + K_2(\varepsilon_+' + \varepsilon_-')$$
$$\varepsilon_+ = K_1\varepsilon_+' + K_2(\varepsilon_+' + \varepsilon_-')$$
$$\varepsilon_- = K_1\varepsilon_-' + K_2(\varepsilon_+' + \varepsilon_-') \tag{16.15}$$

where

$$K_1 = \frac{B'}{B} \quad K_2 = \frac{A'B - AB'}{2AB} \quad A' = \frac{1-\nu}{2} \quad B' = \frac{1+\nu}{2} \tag{16.16}$$

ε_0', ε_+', ε_-': are the measured (relieved) strains at angles θ, $(\theta + 45°)$, $(\theta - 45°)$, respectively.

ε_0, ε_+, ε_-: are the residual strains at angles θ, $(\theta + 45°)$, $(\theta - 45°)$, respectively.

Equation (16.15) determines the residual strains in terms of the measured strains. From the residual strains, the residual stresses are calculated by applying Hooke's law.

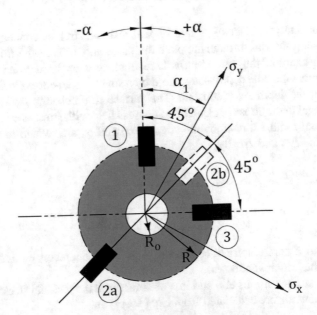

Fig. 16.4 A three-element strain gage rosette for hole-drilling method

Fig. 16.5 A three-element strain-gage rosette for hole-drilling (courtesy of Micro-Measurements Inc.)

16.5 Variation of Residual Stresses Through the Thickness

The previous analysis of uniaxial and biaxial residual stresses concerned with the case of constant stresses through the plate thickness and through-the-thickness hole. We will now examine the case of variable residual stresses through the thickness.

Consider that the hole depth is increased by a small increment Δz and the average stress σ over the depth Δz is constant. The surface strain change $\Delta \varepsilon$ is assumed to be proportional to the stress σ. For a biaxial stress field with stresses σ_l and σ_t along the orthogonal l and t directions the stresses σ_l and σ_t are related to the relieved surface strains $\Delta \varepsilon_l$ and $\Delta \varepsilon_t$ by

$$\Delta \varepsilon_l = \frac{1}{E}(K_1 \sigma_l - \nu K_2 \sigma_t)$$

$$\Delta \varepsilon_t = \frac{1}{E}(K_1 \sigma_t - \nu K_2 \sigma_l) \tag{16.17}$$

where K_1 and K_2 are empirical constants. Note that the stresses σ_l and σ_t are not necessarily the principal stresses.

When the surface strains $\Delta \varepsilon_l$ and $\Delta \varepsilon_t$ are measured the residual stresses over the depth increment Δz are calculated from Eq. (16.17) as

$$\sigma_l = \frac{E}{K_1^2 - \nu K_2^2}[K_1(\Delta \varepsilon_l) + \nu K_2(\Delta \varepsilon_t)]$$

$$\sigma_t = \frac{E}{K_1^2 - \nu K_2^2}[K_1(\Delta\varepsilon_t) + \nu K_2(\Delta\varepsilon_l)] \qquad (16.18)$$

If the l and t are not the principal directions, the surface strains must be recorded in at least three directions. The principal subsurface stresses and their directions are then evaluated using the rosette equations together with Hooke's law.

The validity of Eqs. (16.17) was demonstrated experimentally with K_1 and K_2 empirical proportionality constants. Using Eqs. (16.17), K_1 and K_2 are related to the subsurface stresses averaged over the increment and the measured surface strain increments by

$$K_1 = \frac{E}{\sigma_l^2 - \sigma_t^2}[\sigma_l(\Delta\varepsilon_l) - \sigma_t(\Delta\varepsilon_t)]$$

$$K_1 = \frac{-E}{\nu(\sigma_l^2 - \sigma_t^2)}[\sigma_l(\Delta\varepsilon_t) - \sigma_t(\Delta\varepsilon_l)] \qquad (16.19)$$

K_1 and K_2 are obtained from a supplementary calibration test of known stresses, like a uniaxial tension test by drilling a hole and recording the surface strain changes. When K_1 and K_2 have known the residual stresses for the unknown problem can be evaluated by Eqs. (16.18).

The above method of evaluating residual stresses was developed by Kelsey [1]. The method is effective for stresses up to a depth from the surface equal to about half the hole diameter.

16.6 Nondestructive Methods for Measuring Residual Stresses

The most well developed nondestructive methods for measuring residual stresses are X-ray, neutron diffraction, ultrasonic and magnetic methods. These methods were briefly discussed in Chap. 15.

Further Readings

1. Kula E (1982) Residual stress and stress relaxation. Springer
2. Noyan IC, Cohen JB (1987) Residual stress: measurement by diffraction and interpretation. Springer
3. Dally JW, Riley WF (1991) Experimental stress analysis, 3rd ed. McGraw-Hill, pp 329–334
4. Hutchings MT, Krawitz AD (eds) (1992) Measurement of residual and applied stress using neutron diffraction. Springer Science + Business Media
5. Rowlands RE (1993) Residual stresses. In: Kobayashi AS (ed) Handbook on experimental mechanics, 2nd ed. Society for Experimental Mechanics, pp 785–824

6. Lu J (ed) (1996) Handbook of measurement of residual stresses. Society for Experimental Mechanics Inc., Fairmont Press, Lilburn
7. Totten G, Howes M, Inoue T (2002) Handbook of residual stress and deformation of steel. ASM International
8. Youtsos A (ed) (2006) Residual stress and its effects on fatigue and fracture. Springer
9. Cheng W, Finnie I (2007) Residual stress measurement and the slitting method. Springer
10. Kudryavtsev YF (2008) Residual stress. In: Sharpe WN (ed) Handbook of experimental solid mechanics. Springer, pp 371–387
11. ASTN, E837–08 (2009) Test method for determining residual stresses by the hole-drilling strain-gage method. ASTM International
12. Razumovsky IA (2011) Interference-optical methods of solid mechanics. Springer, pp 125–155
13. Proulx T (2011) Engineering applications of residual stress, vol 8. Springer
14. Schöbel M (2012) Residual stresses and thermal fatigue in metal matrix composites. Südwestdeutscher Verlag für Hochschulschriften
15. Hutchings MT, Withers PJ, Holden TM, Lorentzen T (2005) Introduction to characterization of residual stress by neutron diffraction. CRC Press
16. Schajer GS (ed) (2013) Practical residual stress measurement methods. Wiley
17. Niku-Lari A (ed) (2013) Residual stresses. Pergamon
18. Rossi M, Sasso M, Connesson N, Singh R, DeWald A, Backman D, Gloeckner P (2013) Residual stress, thermomechanics & infrared imaging, hybrid techniques and inverse problems, vol 8. Springer
19. Shukla A, Dally JW (2014) Experimental solid mechanics, 2nd ed. College House Enterprises, pp 244–251
20. Shokrieh MM (ed) (2014) Residual stresses in composite materials. Woodhead Publishing
21. Bossuyt S, Schajer G, Carpinteri A (eds) (2016) Residual stress, thermomechanics & infrared imaging, hybrid techniques and inverse problems, vol 9. Springer
22. Quinn S, Balandraud X (2017) Residual stress, thermomechanics & infrared imaging, hybrid techniques and inverse problems, vol 9. Springer
23. Schajer GS, Whitehead FS (2018) Hole-drilling method for measuring residual stresses. Morgan & Claypool Publishers
24. Baldi A, Quinn S, Balandraud X, Dulieu-Barton JM, Bossuyt S (eds) (2018) Residual stress, thermomechanics & infrared imaging, hybrid techniques and inverse problems, vol 7. Springer
25. Wang H, Zhu L, Xu B (2018) Residual stresses and nanoindentation testing of films and coatings. Springer
26. Baldi A, Kramer SLB, Pierron F, Considine J, Bossuyt S, Hoefnagels J (eds) (2020) Residual stress, thermomechanics & infrared imaging and inverse problems, vol 6. Springer
27. Schajer GS (1988) Measurement of non-uniform residual stresses using the hole-drilling method. Pat I - stress calculation procedures. J Engng Mat Trans ASME 110:338–343
28. Withers PJ, Bhadeshia HKDH (2001) Overview. Residual stress: Part 1—measurement techniques. Mat Sci Tech 17:355–365
29. Rossini NS, Dassisti M, Benyounis KY, Olabi AG (2012) Review: methods of measuring residual stress in components. Mater Des 35:572–588

Index

© The Editor(s) (if applicable) and The Author(s), under exclusive license
to Springer Nature Switzerland AG 2022
E. E. Gdoutos, *Experimental Mechanics*, Solid Mechanics and Its Applications 269,
https://doi.org/10.1007/978-3-030-89466-5